Progress in Automation and Information Systems

Progress in Automation and Information Systems

J.S. Baras and V. Mirelli (eds.): Recent Advances in Stochastic Calculus
J.S. Baras and V. Mirelli (eds.): Recent Advances in Stochastic Systems

John S. Baras Vincent Mirelli
Editors

Recent Advances in Stochastic Calculus

Springer-Verlag
New York Berlin Heidelberg
London Paris Tokyo Hong Kong

John S. Baras
University of Maryland
College Park, MD 20742

Vincent Mirelli
Center for Night Vision and Electro-Optics
Fort Belvoir, VA 22060

Managing Editor

John S. Baras
University of Maryland
College Park, MD 20742

Library of Congress Cataloging-in-Publication Data
Recent advances in stochastic calculus / John S. Baras, Vincent Mirelli, editors.
p. cm.—(Progress in automation and information systems ; 1)
"Includes the material presented in the Distinguished lecture series on stochastic calculus at the Systems Research Center of the University of Maryland at College Park in 1987"—Pref.

1. Stochastic analysis. I. Baras, John S. II. Mirelli, Vincent. III. Title: Distinguished lecture series on stochastic calculus. IV. Series.
QA274.2.R43 1990
519.2—dc20 90-33059

Printed on acid-free paper

Softcover reprint of the hardcover 1st edition 1990

Camera-ready copy prepared from the editors' TeX file.

9 8 7 6 5 4 3 2 1

ISBN-13: 978-1-4612-7999-0 **e-ISBN-13: 978-1-4612-3408-1**
DOI: 10.1007/978-1-4612-3408-1

PREFACE

This volume includes the material presented in the *Distinguished Lecture Series on Stochastic Calculus* at the Systems Research Center of the University of Maryland at College Park in 1987. The purpose of these lecture series and the volume is to acquaint a wide audience with certain recent advances in stochastic calculus and with their applications to significant problems.

Stochastic systems play a fundamental role in automation and information engineering. The analysis of stochastic systems depends in a fundamental way on stochastic calculus. The subject matter is rather sophisticated, requiring a broad mathematical sophistication and maturity. Yet improper understanding or utilization of stochastic calculus in applications, especially in engineering, can lead to incorrect numerial results and faulty designs.

The material included in this volume appears for the first time in book form. Considerable effort was undertaken by the authors to present the material in a form accessible to as wide an audience as possible. Some of the material appears here for the first time, while we believe that the targeted tutorial and survey nature of some of the chapters should be extremely helpful to researchers studying recent developments in stochastic calculus. The topics were selected to cover some of the most important areas for stochastic control, stochastic filtering and stochastic modeling.

Robert Elliott, in chapter 1, provides an exposition of a unified treatment of filtering smoothing and prediction problems when the observation process is a point process whose intensity is a function of the signal. These results are then used to provide an elegant treatment of the stochastic control problem of a partially observed Markov chain with point process observations.

The treatment is based on the derivation of the unnormalized Zakai equations, which are derived for the three situations: filtering, smoothing and prediction. The stochastic control problem is treated in separated form by considering the control of a related Zakai equation. As a result, a minimum principle is obtained. Finally, an approximate minimum principle satisfied by an almost optimal control is derived using a variational inequality.

Alain Bensoussan, in chapter 2, presents a unified treatment of nonlinear filtering theory. The treatment of this problem provided in this chapter is focused on diffusion signal and observation models. The main core of theory presented consists in characterizing conditional expectations of functions of the state process, as solutions of an evolution equation. Both normalized and unnormalized conditional expectations are considered. The corresponding evolution equations are known as Kushner-Stratonovitch and Zakai equations respectively. Existence and uniqueness of solutions in adequate spaces is proved in this chapter for both types of evolution equations. In addition, both cases of correlated and uncorrelated signal and observation noises are considered. Subsequently, when conditional densities exist, the corresponding stochastic partial differential equations are studied, and via the so called robust from, explicit representation formulas for the densities are developed. These representations are extensions of the well known probabilistic interpretations of solutions of parabolic equations.

Dominique Michel and Etienne Pardoux, in chapter 3, provide a comprehensive treatment of Malliavin calculus and its applications. This is a recently developed stochastic calculus of variation which is in particular capable of giving powerful criteria for the law of a given functional of Brownian motion to possess a density. Starting from the simpler case of a finite dimensional probability space, the ideas of Malliavin and Bismut are presented. This introduction has important educational significance. Subsequently, the basic definitions and results of the Malliavin Calculus on Wiener space are developed following Bismut and Zakai. The sufficient condition for the existence of a density is linked to the Hörmander condition. Malliavin Calculus is also applied to the problem of existence of densities for the conditional law in nonlinear filtering as well as to the existence of finite dimensional sufficient statistics. Finally, it is shown how the differential calculus on Wiener space can be used to generalize Itô's and Stratonovitch's stochastic integrals and differential calculus.

Tyrone Duncan, in chapter 4, provides an exposition of the foundations of stochastic calculus on manifolds. This includes the construction of Brownian motion and its horizontal lift to the bundle of orthoromal frames for a compact Riemannian manifold. Extensions, to noncompact manifolds are discussed. Subsequently, a wide range of applications of stochastic calculus on manifolds is described. These include a local formula for an index theorem, a result on the nonexistence of harmonic 1-forms and the description of a theta function for a compact Lie algebra as an integral of Wiener measure on a compact Lie group. Applications to probability given include an explicit expression for the Radon-Nikodym derivative for a measure absolutely continuous with respect to Wiener measure in a manifold and a law of the iterated logarithm for a manifold-valued Brownian motion. Some applications to stochastic control and filtering are given including the existence of an optimal control for a controlled diffusion process on a manifold satisfying a convexity condition and necessary and sufficient conditions for an optimal control of a partially observed control diffusion process in a manifold. The nonlinear filtering problem for observation processes on manifolds is also treated, and two families of explicitly solvable problems are described. The first family involves stochastic systems in real hyperbolic spaces, while the second involves stochastic systems on spheres.

Vivek Borkar, in chapter 5, provides a survey on some recent results on controlled Markov chains. The main idea in the approach is to pose the control problem as an optimization problem on an appropriate set of probability measures. For cost criteria such as discounted cost, finite time horizon cost and cost up to an exit time, this approach not only recovers all known results from a different perspective, but in addition it appears promising for constrained optimization problems as well. For the long-run average cost, this approach leads to general results which subsume most known results on this problem. In this case the stability-based classification of some of the control strategies introduced is also provided.

Thomas Kurtz, in chapter 6, provides a penetrating investigation of the martingale problem for constrained processes, i.e. for processes constrained to remain in a closed set of a metric space. The analysis leads to constructive methods which involve the introduction of a "patchwork martingale problem." Based on the solution of the latter, the constrained martingale problem is analyzed. It is shown that a local time solution of a constrained martingale problem can be constructed from a solution of the corresponding patchwork martingale problem provided λ_o is strictly increasing and tends to infinity. The converse also holds. Here λ_o is the local time associated with the set where the processes is constrained. In addition to existence of solutions, the question of uniqueness is also examined. For reflecting diffusions a useful notion is that of a submartingale problem. It is established, under quite general conditions, that a solution of the submartingale problem is also a solution of the constrained martingale problem. Finally, it is explained how the so called "penalty method" for constructing reflecting diffusion processes fits into the framework of constrained martingale problems.

Ioannis Karatzas, in chapter 7, presents a unified approach, based on stochastic calculus, to problems of option pricing and consumption/investment, under very general assumptions on the market model. The financial market considered includes a risk-free asset, called the bond, and several risky assets called stocks. Stock prices are driven by independent Brownian motions which model the exogenous forces of uncertainty that influence the market. The interest rate of the bond, the appreciation rates of the stocks and their volatilities are modeled as arbitrary bounded measurable processes adopted to the Brownian filtration. The methodologies used are based on stochastic calculus, such as the Girsanov change of probability measure and the representation of Brownian martingales as stochastic integrals. A general treatment of the pricing of contingent claims such as options (European and American) is provided. The resolution of consumption/investment problems for a "small investor" (i.e. an economic agent whose actions cannot influence the market prices) with quite general utility functions is given as well. Finally, equilibrium models are studied in detail. These are formulated in the context of an economy with several small investors and one consumption good, whose price is determined by the joint optimal actions of all these agents in a way that "clears" the markets.

These lecture series provided a unique environment for exchange of ideas and strong interactions at the Systems Research Center during 1987. We are grateful to the authors and many colleagues and students for their participation and contributions. The cross disciplinary environment of the Center was instrumental in the overall organization, development and success of the series. We sincerely hope that this volume will transmit the excitement and better understanding achieved to the community at large. The support from the Unites States Army Research Office and the National Science Foundation was vital to the undertaking of the series. Finally we would like to express our sincere gratitude to Lien Trieu, Linda Britt, Cherri Helms, and Helen Treherne for typsetting and producing the final form of the manuscript. We would also like to thank Pamela Harris, Sharon Dass and Maggie Virkus for efficiently managing the overall process.

John S. Baras
College Park, Maryland

Vincent Mirelli
Fort Belvior, Virginia

CONTENTS

1
FILTERING AND CONTROL FOR POINT PROCESS OBSERVATIONS

Robert J. Elliott

INTRODUCTION

The following notes discuss the filtering, smoothing and prediction problems when the observation process is a point process whose intensity is a function of the signal. The linear, unnormalized Zakai equations are derived in the three situations. In section 1.5 the control of a partially observed Markov chain is treated in separated form by considering the control of a related Zakai equation, and a minimum principle obtained. Finally, an approximate minimum principle satisfied by an almost optimal control is derived using a variational inequality of Ekeland, [3].

These lectures were presented at the Systems Research Center at the University of Maryland in the Spring of 1986. The author would like to thank Dr. Baras, Director of the Systems Research Center, for the invitation, and the Center and the Department of Mathematics of the University of Maryland for their hospitality.

1.1 FILTERING WITH POINT PROCESS OBSERVATIONS

All processes are defined on a probability space $(\Omega, \underline{F}, P)$ which has a filtration $\underline{\underline{F}} = \{\underline{F}_t\}, t \geq 0,$ to which they are all adapted. It is supposed the filtration satisfies the 'usual conditions,' that is $\underline{\underline{F}}$ is complete, so each $\underline{F}_t$ contains all subsets of all null sets of $\underline{F}$, and the filtration is right continuous:

$$\underline{F}_t = \underline{F}_{t+} = \bigcap_{s>t} \underline{F}_s.$$

The **signal process** will be a (real) semimartingale

$$\xi_t = \xi_o + \int_o^t \beta_u \, du + M_t, \qquad t \geq 0.$$

For simplicity we assume $E[\xi_o^2] < \infty, E[\int_o^\infty \beta_u^2 \, du] < \infty$ and M is a square integrable $\underline{\underline{F}}$ martingale.

The **observation process** will be a point process $N = \{N_t\}, t \geq 0$, with an intensity $h(u, \xi_u) = h(u)$. Consequently, $h(u)$ is a non-negative process, and again we shall assume $E[\int_o^\infty h(u)^2 du] < \infty$. If the trajectory $\xi_u(\omega), u \geq 0$, is known N_t is a conditional Poisson process with intensity $h(u)$, and

$$Q_t = N_t - \int_o^t h(u) \, du$$

is an $\underline{\underline{F}}$ martingale.

Write $\underline{G}_t^o = \sigma\{N_s : s \leq t\}$, and $\underline{\underline{G}} = \{\underline{G}_t\}$ for the right continuous, complete filtration generated by the $\underline{G}_t^o$. Then for all $t \geq 0$ $\underline{G}_t \subset \underline{F}_t$.

If $\{\phi_t\}$, $t \geq 0$, is any process write $\Pi(\phi_t)$ or $\hat{\phi}_t$ for the $\underline{\underline{G}}$ optional projection of ϕ. Now, (see [2], p.60),

$$\pi(\phi_t) = \hat{\phi}_t = E[\phi_t \mid \underline{G}_t] \quad \text{a.s.} \tag{1.1.1}$$

for each $t \geq 0$, (but this equation cannot be used to define the optional projection because it takes no account of measurability in t.) Similarly, write $\hat{\hat{\phi}}$ for the $\underline{\underline{G}}$ predictable projection of ϕ. Again,

$$\hat{\hat{\phi}}_t = E[\phi_t \mid \underline{G}_{t-}] \quad \text{a.s.} \tag{1.1.2}$$

Lemma 1.1 For any process ϕ

$$\int_o^t \hat{\phi}_u \, du = \int_o^t \hat{\hat{\phi}}_u \, du \ \text{ a.s., for all } t > 0.$$

Proof: For any $\underline{\underline{G}}$ stopping time T the process $X = I_{]]o,T\wedge t]]}$ is $\underline{\underline{G}}$ predictable. From Theorems 6.39 and 7.25 of Elliott [4],

$$\begin{aligned} E[\int_o^\infty \Pi_p(X\hat{\phi})_u \, du] &= E[\int_o^\infty X_u \, \hat{\phi}_u \, du] \\ &= E[\int_o^{T\wedge t} \hat{\phi}_u \, du] = E[\int_o^{T\wedge t} \Pi_p(\phi)_u \, du] \\ &= E[\int_o^{T\wedge t} \hat{\hat{\phi}}_u \, du]. \end{aligned}$$

Therefore, the process $\Psi_s = \int_o^{s\wedge t} \hat{\phi}_u \, du - \int_o^{s\wedge t} \hat{\hat{\phi}}_u \, du$ is a $\underline{\underline{G}}$ martingale, by 4.18 of [4]. However, it is predictable and of bounded variation, and so zero. (Theorem 11.39 of [4]. This result is also an immediate consequence of the fact that for each ω $\hat{\phi}_t = \hat{\hat{\phi}}_t$, except for possibly a countable number of t values.)

Notation 1.2 Write $\widetilde{Q}_t = N_t - \int_o^t \hat{\hat{h}}(u) \, du$.

Lemma 1.3 $\widetilde{Q}$ is a $\underline{\underline{G}}$ martingale.

Proof:

$$E[\widetilde{Q}_t - \widetilde{Q}_s | \underline{\underline{G}}_s] = E[Q_t - Q_s + \int_s^t (h(u) - \hat{\hat{h}}(u)) \, du | \underline{\underline{G}}_s].$$

Using Lemma 1.1. this is

$$=E\,[E\,[Q_t - Q_s | \, \underline{\underline{F}}_s]\, \underline{\underline{G}}_s] + E\,[\int_s^t h(u) - \hat{h}(u) \, du | \, \underline{\underline{G}}_s]$$

$$=0,$$

by Fubini's Theorem and (1).

Remark 1.4 The objective of filtering theory is to obtain a recursive expression for the $\underline{\underline{G}}$ optional projection $\hat{\xi}_t$ of the signal process. Calculations similar to those of Lemma 1.3 show that the process $\widetilde{M}_t = \hat{\xi}_t - \hat{\xi}_o - \int_o^t \hat{\beta}_u \, du$ is a square integrable $\underline{\underline{G}}$ martingale, and so $\widetilde{M}_t$ can be represented as a stochastic integral

$$\widetilde{M}_t = \int_o^t \gamma_u \, d\, \widetilde{Q}_u$$

where $E\,[\int_o^\infty \gamma_u^2 \hat{\hat{h}} \, du] < \infty$.

(The proof of the representation result can be reduced to considering one jump at a time of the point process. Suppose $\{T_n\}$ are the jump times of N. Then $\widetilde{M}_t = \sum_n \Delta \widetilde{M}_{T_n \wedge t}$ where, by the optional stopping theorem, the $\Delta \widetilde{M}_{T_n \wedge t} = \widetilde{M}_{T_{n+1} \wedge t} - \widetilde{M}_{T_n \wedge t}$ are orthogonal, square integrable martingales.) The problem is to find an explicit formula for the integrand γ.

Theorem 1.6

$$\gamma_u = I_{(\hat{\hat{h}}_u \neq 0)} (\hat{\hat{h}}(u))^{-1} (\widehat{\widehat{\xi_{u-} h_u}} - \hat{\xi}_{u-} \hat{\hat{h}}_u + \frac{d}{du} < \widehat{\widehat{M, N}} >).$$

Proof: By the product rule

$$\xi_t N_t = \int_o^t \xi_{u-} (d\, Q_u + h_u \, du) + \int_o^t N_{u-} (\beta_u \, du + dM_u)$$
$$+ \sum_{0<u\leq t} \Delta \xi_u \, \Delta \, N_u. \tag{1.1.13}$$

Now

$$\sum_{0<u\leq t} \Delta\xi_u \,\Delta N_u = \sum_{0<u\leq t} \Delta M_u \,\Delta N_u$$
$$=[M,\, N].$$

Suppose the $\underline{\underline{F}}$ dual predictable projection $< M,\; N >$ of this process exists. Taking the $\underline{\underline{G}}$ optional projections of the processes in (1.1.3), calculations similar to Lemma 1.3 show that H^1 is a locally square integrable $\underline{\underline{G}}$ martingale where

$$H^1_t = \widehat{\xi_t N_t} - \int_o^t \widehat{\xi_{u-} h_u}\, du - \int_o^t \widehat{N_{u-}\beta_u}\, du - < \widehat{M, N} >_t . \tag{1.1.4}$$

Note, therefore, that $\widehat{\xi_t N_t} = \hat{\xi}_t N_t = N_t E\left[\xi_t | \underline{\underline{G}}_t\right]$ is a **special semimartingale**, so its decomposition, as the sum of a martingale and a predictable bounded variation process, is unique. However, we also have

$$\begin{aligned}\hat{\xi}_t N_t = &\int_o^t \hat{\xi_{u-}}(d\widehat{Q}_u + \hat{\hat{h}}_u\, du)\\ &+ \int_o^t N_{u-}(\hat{\beta}_u\, du + \gamma_u\, d\widetilde{Q}_u) + \sum_{0<u\leq t} \Delta\hat{\xi}_u \,\Delta N_u .\end{aligned}$$

Now $\Delta\hat{\xi}_u = \gamma_u \,\Delta N_u$, so the last sum is

$$\begin{aligned}\sum_{0<u\leq t} \gamma_u(\Delta N_u)^2 &= \sum_{0<u\leq t} \gamma_u \,\Delta N_u = \int_o^t \gamma_u\, dN_u\\ &= \int_o^t \gamma_u\, d\widetilde{Q}_u + \int_o^t \gamma_u\, \hat{\hat{h}}_u\, du.\end{aligned}$$

Then H^2 is a locally square integrable $\underline{\underline{G}}$ martingale, where

$$\begin{aligned}H^2_t = \hat{\xi}_t\, N_t &- \int_o^t \hat{\xi}_{u-}\, \hat{\hat{h}}_u\, du - \int_o^t N_{u-}\, \hat{\beta}_u\, du\\ &- \int_o^t \gamma_u \hat{\hat{h}}_u\, du.\end{aligned} \tag{1.1.5}$$

Equating the bounded variation processes in the two representations (1.1.4), (1.1.5) of the special semimartingale $\hat{\xi}_t \;N_t$

$$\gamma_u \hat{\hat{h}}_u = \widehat{\xi_{u-} h_u} - \hat{\xi}_{u-}\, \hat{\hat{h}}_u + \frac{d}{du} < \widehat{M, N} >_u$$

giving

$$\gamma_u = (\hat{\hat{h}}_u)^{-1}(\widehat{\xi_{u-} h_u} - \hat{\xi}_{u-}\, \hat{\hat{h}}_u + \frac{d}{du} < \widehat{M, N} >_u)$$

if $\hat{\hat{h}}_u \neq 0$.

For any set $A \in \underline{G}_s$

$$E\left[I_A \int_s^t dN_u\right] = E\left[I_A \int_s^t \hat{\hat{h}}_u \, du\right]$$

so γ_u can be taken as 0 on any set where $\hat{\hat{h}}_u = 0$.

Remarks 1.7 The random variables γ_u are $\underline{G}_{u-}$ measurable. If M and N have no jumps in common $[M, N] = 0$, so $< M, N >= 0$.

Example 1.8 An application of the above result is to the so-called 'disorder problem.' Suppose $T > 0$ is a failure time with an absolutely continuous distribution function $G_t = P(T \leq t) = \int_o^t f(u)\, du$.

Write $F_t = 1 - G_t$, and consider the processes $\xi_t = I_{t \geq T}$, and

$$M_t = I_{t \geq T} - \int_o^t \frac{(1-\xi_u)\, f_u \, du}{F_u}.$$

Then M_t is a martingale with respect to the filtration $\underline{\underline{F}}$ generated by ξ, so the Doob-Meyer decomposition of ξ is

$$\xi_t = \int_o^t \frac{(1-\xi_u)\, f_u}{F_u} du + M_t.$$

Suppose at time T the intensity of the point process N changes from a to b. Then the intensity of N is

$$h_u = a + (b-a)\, \xi_u$$

and

$$\hat{\hat{h}}_u = a + (b-a)\, \hat{\xi}_{u-}$$

The jump time T is, almost surely, not a jump of N, so the filtering formula of Theorem 1.6 gives

$$\hat{\xi}_t = \int_o^t \frac{(1-\hat{\xi}_u) f_u du}{F_u} + \int_o^t \frac{(b-a)(\hat{\xi}_{u-} - (\hat{\xi}_{u-})^2)\, d\widetilde{Q}_u}{a + (b-a)\hat{\xi}_{u-}}.$$

Example 1.9 Consider a finite state Markov process X with state $e_1, \ldots, e_N$, where the e_i are the unit column vectors in R^N. Write $p_t^i = P(X_t = e_i)$ and suppose the probability column vector $p_t = (p_t^1, \dots, p_t^N)'$ satisfies the Kolmogorov forward equation $\frac{dp}{dt} = A\, p_t$, where $A = (a^{ij})$ is a constant generator. Then

$$X_t = X_o + \int_o^t A\, X_u\, du + M_t$$

where M is martingale. Suppose N is a conditional Poisson process, with intensity $h(X_u)$, so as above we can write

$$\begin{aligned} N_t &= \int_o^t h(X_u)du + Q_t \\ &= \int_o^t \hat{\hat{h}}(X_u)du + \widetilde{Q}_t. \end{aligned}$$

Write $\hat{p}_u^i = P(X_u = e_i \mid \underline{G}_u)$. Then

$$\begin{aligned} \hat{\hat{h}}(X_u) &= E[h(X_u) \mid \underline{G}_{u-}] \\ &= \sum_{i=1}^N h(e_i)\, \hat{p}_{u-}^i. \end{aligned}$$

Write B for the diagonal matrix with entries $h(e_i)$. Then the filtering formula of Theorem 1.6 gives

$$\hat{p}_t = \hat{p}_o + \int_o^t A\hat{p}_u\, du + \int_o^t I_{(\hat{\hat{h}}_u \neq 0)}(\hat{\hat{h}}_u)^{-1}(B\hat{p}_{u-} - \hat{\hat{h}}_u\, \hat{p}_{u-})d\widetilde{Q}_u.$$

(Note that in (1.1.6)

$$\begin{aligned} \hat{\hat{h}}(X_s) &= E[h(X_s) \mid \underline{G}_{s-}] \\ &= \sum_{i=1}^N h(e_i)P(X_s = e_i \mid \underline{G}_{s-}) \end{aligned}$$

and

$$P(X_x = e_i \mid \underline{G}_{s-}) = P(X_{s-} = e_i \mid \underline{G}_{s-}) = \hat{p}_{s-}^i$$

because

$$\begin{aligned} &E[X_s - X_{s-} \mid \underline{G}_{s-}] \\ &= E[E[M_s - M_{s-} \mid \underline{F}_{s-} \mid \underline{G}_{s-}] = 0.) \end{aligned}$$

1.2 THE REFERENCE PROBABILTY AND DETECTION

The signal process will again be a semimartingale ξ as in section 1.1, and the observation process will be a point process N. Suppose that under a probability measure P_1, (the reference probability), N is a standard Poisson process of intensity $h_t = 1$. Then

$$\overline{Q}_t = N_t - t \tag{1.2.7}$$

is a $\underline{\underline{G}}$ martingale under P_1. Suppose that $h(u, \xi_u) = h_u$ is as above and consider the family of exponentials

$$\Lambda_t = \Pi_{0<u\leq t} h(u, \xi_u) \triangle N_u \ \exp\left(\int_o^t (1 - h(u, \xi_u))\, du\right).$$

Then, (see [4] p. 156), Λ is an $(\underline{\underline{F}}, P_1)$ martingale and

$$\Lambda_t = 1 + \int_o^t \Lambda_{u-}(h_u - 1)(dN_u - du). \tag{1.2.8}$$

Suppose a new probability measure $P = P_h$ is defined by

$$E\,[\frac{dP}{dP_1} \mid F_t] = \Lambda_t.$$

The following point-process form of Girsanov's theorem is then true:

Theorem 1.2.1 Under P the process

$$Q_t = N_t - \int_o^t h_u\, du \tag{1.2.9}$$

is an $\underline{\underline{F}}$ martingale.

Proof: By the Baye's formula (see Loève [8], §27.4)

$$\begin{aligned} E\,[Q_t | \underline{\underline{F}}_s] &= \frac{E_1\,[\Lambda_t Q_t | \underline{\underline{F}}_s]}{E_1\,[\Lambda_t | \underline{\underline{F}}_s]} \\ &= \Lambda_s^{-1} E_1[\Lambda_t\, Q_t | \,\underline{\underline{F}}_s], \end{aligned}$$

where E (resp. E_1) denotes expectation with respect to P (resp. P_1). Clearly Q_t is a $P-$ martingale if, and only if, $\Lambda_t\, Q_t$ is a P_1- martingale.

Now

$$\Lambda_t Q_t = \int_o^t \Lambda_{u-}\, dQ_u + \int_o^t Q_{u-}\, d\Lambda_u + [\Lambda, Q]_t.$$

However,

$$[\Lambda, Q]_t = \sum_{0<u\leq t} \Delta\Lambda_u \, \Delta Q_u = \int_o^t \Lambda_{u-}(h_u - 1) d N_u.$$

Therefore,

$$\Lambda_t Q_t = \int_o^t \Lambda_{u-} \, h_u \, d\overline{Q}_u + \int_o^t Q_{u-}\Lambda_{u-}(h_u - 1) d\overline{Q}_u.$$

Because $\overline{Q}_t = N_t - t$ is a martingale under P_1, the result follows.

Remarks 1.2.2 Under P the point process N is a conditional Poisson process with intensity h_u. If N is standard Poisson $N_t - t$ is a martingale, whilst if N is a conditional Poisson with rate $h_u, N_t - \int_o^t h_u \, du$ is a martingale. Depending which is the case, different probability measures are induced on the space of point process. The problem is then investigated by considering the Radon-Nikodym derivative Λ_t. Alternatively, the filtering problem can be formulated in terms of P_1, P and Λ.

Notation 1.2.3 Write $\hat{\Lambda}$ for the $\underline{\underline{G}}$ optional projection of Λ under P_1. Then for each $t \geq 0$

$$\hat{\Lambda}_t = E_1[\Lambda_t | \underline{\underline{G}}_t].$$

(N.B. the expectation here is with respect to P_1).

Then $\hat{\Lambda}$ is a locally square integrable $(\underline{\underline{G}}, P_1)$ martingale and so it has a stochastic integral representation with respect to $\bar{Q}$

$$\hat{\Lambda}_t = 1 + \int_o^t \lambda_u \, d\,\bar{Q}_u. \tag{1.2.10}$$

In this section the $(\underline{\underline{G}}, P)$ predictable projection of h_u will be denoted by

$$\Pi_p(h_u),$$

so this equals $E[h_u | G_{u-}]$ a.s.

The filtering and detection result is then:

Theorem 1.2.4

$$\lambda_u = \hat{\Lambda}_{u-}(\Pi_p(h_u) - 1).$$

i.e.

$$\hat{\Lambda}_t = 1 + \int_o^t \hat{\Lambda}_{u-}(\Pi_p(h_u) - 1) d\,\bar{Q}_u.$$

Proof: By the product rule

$$\Lambda_t N_t = \int_o^t \Lambda_{u-} dN_u + \int_o^t N_{u-}\Lambda_{u-}(h_u - 1)d\bar{Q}_u + [\Lambda, N]_t.$$

Now,

$$\begin{aligned}[\Lambda, N]_t &= \sum_{0<u\leq t} \triangle\Lambda_u \, \triangle N_u \\ &= \int_o^t \Lambda_{u-}(h_u - 1)d N_u.\end{aligned}$$

Therefore,

$$\Lambda_t N_t = \int_o^t \Lambda_{u-} \, h_u \, du + K_t^1,$$

where K^1 is locally square integrable $(\underline{\underline{F}}, P_1)$ martingale. Taking the $\underline{\underline{G}}$ optional projections under P_1, and using Lemma 1.1.

$$\hat{\Lambda}_t N_t = \int_o^t \widehat{\Lambda_{u-} h_u} \, du + \mu_t. \tag{1.2.11}$$

Here $\widehat{\ }$ denotes the $\underline{\underline{G}}$ predictable projection under P_1 and μ_t is a locally square integrable $\underline{\underline{G}}, P_1$ martingale. Again, $\hat{\Lambda}_t N_t$ is a special semimartingale. Using(1.2.10)

$$\hat{\Lambda}_t N_t = \int_o^t \hat{\Lambda}_{u-} d N_u + \int_o^t N_{u-} \, \lambda_u \, d\bar{Q}_u + [\hat{\Lambda}, N]_t.$$

Here,

$$[\hat{\Lambda}, N]_t = \sum_{0<u\leq t} \triangle\hat{\Lambda}_u \, \triangle N_u = \int_o^t \lambda_u dN_u,$$

so

$$\hat{\Lambda}_t N_t = \int_o^t (\hat{\Lambda}_{u-} + \lambda_u) \, ds + \nu_t \tag{1.2.12}$$

where ν_t is a locally square integrable P_1 martingale. The decomposition of special semimartingales is unique, so equating the bounded variation terms in (1.2.11) and (1.2.12)

$$\begin{aligned}\lambda_u &= \widehat{\Lambda_{u-} h_u} - \hat{\Lambda}_{u-} \\ &= E_1[\Lambda_{u-}(h_u - 1)|\underline{G}_{u-}].\end{aligned}$$

Recalling Baye's formula, for any integrable $\underline{F}_t$ measurable, random variable ϕ,

$$\begin{aligned}E[\phi|\underline{G}_t] &= \frac{E_1[\Lambda_t\phi|\underline{G}_t]}{E_1[\Lambda_t|\underline{G}_t]} \\ &= \hat{\Lambda}_t^{-1} E_1[\Lambda_t \phi d|\underline{G}_t],\end{aligned} \tag{1.2.13}$$

we see that

$$\lambda_u = \hat{\Lambda}_{u-}E[(h_u - 1)|\underline{G}_{u-}]$$
$$= \hat{\Lambda}_{u-}(\Pi_p(h_u) - 1)$$

and the result follows.

Remarks 1.2.5 The Zakai equation for point process observations will now be obtained. As in section 1, we wish to discuss the $\underline{\underline{G}}$ optional projection of the signal process ξ under measure P, that is the process

$$\Pi(\xi_t) = E[\xi_t|\underline{G}_t] \quad a.s.$$

For simplicity we shall assume that M and N have no common jumps, so $[M, N] = < M, N >= 0$. Using the Π and Π_p notation for the $(\underline{\underline{G}}, P)$ optional and predictable projections, respectively, Theorem 1.6 states

$$\Pi(\xi_t) = \Pi(\xi_o) + \int_o^t \Pi(\beta_u)\,du + \int_o^t \gamma_u\, d\,\tilde{Q}_u \tag{1.2.14}$$

where

$$\gamma_u = I_{\pi_p(h)\neq 0}(\Pi_p(h_u))^{-1}(\Pi_p(\xi_{u-}\, h_u) - \Pi_p(\xi_{u-})\Pi_p(h_u)).$$

The $\hat{}$ and $\hat{\hat{}}$ notation will denote the $(\underline{\underline{G}}, P_1)$ optional and predictable projections, respectively. Write

$$\sigma(\xi_t) = E_1\,[\xi_t\,\Lambda_t|\underline{G}_t].$$

Then

$$\sigma(1)_t = \hat{\Lambda}_t = 1 + \int_o^t \hat{\Lambda}_{u-}(\Pi_p(h_u) - 1)d\,\bar{Q}_u \tag{1.2.15}$$

and by Baye's formula (1.2.13)

$$\Pi(\xi_t) = E\,[\xi_t|\underline{G}_t] = \frac{\sigma(\xi_t)}{\sigma(1)_t}. \tag{1.2.16}$$

The disadvantage of equation (1.2.14) for $\Pi(\xi_t)$ is that it is quadratic in Π_p. We now derive an equation for $\sigma(\xi_t)$, the Zakai equation, which is linear in the unnormalized distribution σ.

Theorem 1.2.6

$$\sigma(\xi_t) = \sigma(\xi_o) + \int_o^t \sigma(\beta_u)\, du$$
$$+ \int_o^t (\sigma(\xi_{u-}\ h_u) - \sigma(\xi_{u-}))\, d\bar{Q}_u.$$

Proof: From (1.2.16)

$$\sigma(\xi_t) = \hat{\Lambda}_t \Pi(\xi_t),$$

so using (1.2.14) and (1.2.15)

$$\hat{\Lambda}_t \Pi(\xi_t) = \Pi(\xi_o) + \int_o^t \hat{\Lambda}_{u-}\Pi(\beta_u)\, du$$
$$+ \int_o^t \hat{\Lambda}_{u-}\gamma_u\, d\tilde{Q}_u + \int_o^t \Pi(\xi_{u-})\hat{\Lambda}_{u-}(\Pi_p(h_u) - 1)\, d\bar{Q}_u$$
$$+ [\hat{\Lambda}, \Pi(\xi)]_t.$$

Now $\Pi(\xi_o) = \sigma(\xi_o)$, and there are at most countably many points where $\hat{\Lambda}_{u-} \neq \hat{\Lambda}_u$ so

$$\int_o^t \hat{\Lambda}_{u-}(\beta_u)\, du = \int_o^t \hat{\Lambda}_u(\beta_u)\, du$$
$$= \int_o^t \sigma(\beta_u)\, du.$$

Also,

$$[\hat{\Lambda}, \Pi(\xi)]_t = \sum_{0<u\leq t} \Delta\hat{\Lambda}_u\ \Delta\,\Pi(\xi_u)$$
$$= \int_o^t \gamma_u \hat{\Lambda}_{u-}(\Pi_p(h_u) - 1) dN_u.$$

Recalling that in the notation of this section

$$\tilde{Q}_t = N_t - \int_o^t \Pi_p(h_u)\, du$$

we have:

$$\hat{\Lambda}_t \Pi(\xi_t)$$
$$= \sigma(\xi_t) = \sigma(\xi_o) + \int_o^t \sigma(\beta_u)\, du + \int_o^t \hat{\Lambda}_{u-}(\Pi_p(\xi_{u-}\ h_u) - \Pi_p(\xi_{u-})\Pi_p(h_u))d\bar{Q}_u$$
$$+ \int_o^t \hat{\Lambda}_{u-}\Pi(\xi_{u-})(\Pi(h_u) - 1)d\bar{Q}_u$$
$$= \sigma(\xi_o) + \int_o^t \sigma(\beta_u)du + \int_o^t \hat{\Lambda}_{u-}(\Pi_p(\xi_{u-}h_u) - \Pi(\xi_{u-}))d\bar{Q}_u$$
$$= \sigma(\xi_o) + \int_o^t \sigma(\beta_u)du + \int_o^t (\sigma(\xi_{u-}h_u) - \sigma(\xi_{u-}))d\bar{Q}_u.$$

This equation is linear in σ; in addition the $Pi_p(h_u)^{-1}$ term, present in γ_u, has disappeared.

1.3 POINT PROCESS PREDICTION AND SMOOTHING

Again suppose the signal process is a (real) semimartingale

$$\xi_t = \xi_o + \int_o^t \beta_u \, du + M_t, \quad \text{for } 0 \leq t \leq T,$$

with $E[\xi_o^2] < \infty$, $E[\int_o^T \beta_u^2 \, du] < \infty$ and M a square integrable martingale. As in Section 1 the observation process will be a point process with intensity $h(u, \xi_u) = h_u$.

The filtering problem discusses

$$\hat{\xi}_t = E[\xi_t | \underline{G}_t].$$

The smoothing problem discusses

$$\hat{\xi}_{s,t} = E[\xi_s | \underline{G}_t] \text{ for } s \leq t.$$

The prediction problem discusses

$$\hat{\xi}_{t,s} = E[\xi_t | \underline{G}_s] \text{ for } t \geq s.$$

Smoothing. Consider a fixed s with $0 \leq s \leq T$. Then for $t \geq s$ $\hat{\xi}_{s,t}$ is a square integrable $(\underline{\underline{G}}, P)$ martingale, and so it has a representation

$$\hat{\xi}_{s,t} = \hat{\xi}_{s,s} + \int_s^t \gamma_{s,u} \, d\tilde{Q}_u \tag{1.3.17}$$

where $E[\int_o^T \gamma_{s,u}^2 \; \hat{h}_u \, du] < \infty$.

Write $\underline{\underline{H}} = \{\underline{H}_t\}$ for the right-continuous, complete filtration generated by $\underline{G}_t \vee \{\xi_u : 0 \leq u \leq T\}$. Then the intensity $h_u = h(u, \xi_u)$ is $\underline{H}_o$ measurable. Again, for example, $\Pi(\xi_s h_u)$ will denote the $(\underline{\underline{G}}, P)$ optional projection of $\xi_s h_u$, so for all $u \geq s$ $\Pi(\xi_s h_u) = E[\xi_s h_u | \underline{G}_u]$ *a.s.*

Now for $t' \geq t$

$$\begin{aligned} E[\hat{\xi}_{s,t'} N_s | \underline{G}_t] &= E[E[\xi_s N_s | G_{t'}] G_t] \\ &= \hat{\xi}_{s,t} N_s. \end{aligned} \tag{1.2.18}$$

Also, if $D_t = \prod(\int_s^t \xi_s h_u du) - \int_s^t \Pi(\xi_s\ h_u)\ du$ an application of Fubini's theorem, as in Lemma 1.3, shows that

$$E[D_{t'} - D_t | \underline{G}_t] = 0 \text{ for } t' \geq t. \tag{1.2.19}$$

Similarly

$$\begin{aligned} E[E[\xi_s(Q_{t'} - Q_s)|\underline{G}_{t'}]\underline{G}_t] &= E[E[\xi_s(Q_{t'} - Q_s)|\underline{H}_{t'}]\underline{G}_t] \\ &= E[\xi_s E[Q_{t'} - Q_s|\underline{H}_t]\underline{G}_t] \\ &= E[\xi_s(Q_t - Q_s)|\underline{G}_t]. \end{aligned} \tag{1.2.20}$$

From Corollary 6.44 of [4]

$$\Pi(\xi_s N_t) = \hat{\xi}_{s,t} N_t. \tag{1.3.21}$$

Trivially

$$\xi_s N_t = \xi_s N_s + \int_s^t \xi_s\ h_u\ du + \xi_s(Q_t - Q_s). \tag{1.3.22}$$

Taking the $(\underline{G}, P)$ optional projection of both sides of (1.3.22), and using (1.2.18), (1.2.19), (1.2.20) and (1.3.21)

$$\hat{\xi}_{s,t} N_t = \int_s^t \Pi(\xi_s h_u) du + H_t^1 \tag{1.3.23}$$

where H_t^1 is a locally square integrable $(\underline{G}, P)$ martingale. Consequently, $\hat{\xi}_{s,t} N_t$ is a special semimartingale with a unique decomposition. However, from (1.2.18) using the product rule

$$\hat{\xi}_{s,t} N_t = \hat{\xi}_{s,s} N_s + \int_s^t \hat{\xi}_{s,u-} dN_u + \int_s^t N_{u-} \gamma_{s,u} d\tilde{Q}_u + [\hat{\xi}, N]_{s,t}.$$

Now $[\hat{\xi}, N]_{s,t} = \int_s^t \gamma_{s,u} dN_u$.

So

$$\hat{\xi}_{s,t} N_t = \hat{\xi}_{s,s} N_s + \int_s^t (\hat{\xi}_{s,u-} \hat{\hat{h}}_u + \gamma_{s,u} \hat{\hat{h}}_u) du + H_t^2, \tag{1.3.24}$$

where H_t^2 is a locally square integrable $(\underline{G}, P)$ martingale. From (1.3.23) and (1.3.24), using the unique decomposition of special semimartingales:

$$\gamma_{s,u} = I_{(\hat{\hat{h}}_u \neq 0)} (\hat{\hat{h}}_u)^{-1} (\Pi(\xi_s h_u) - \hat{\xi}_{s,u-} \hat{\hat{h}}_u).$$

Therefore, the following smoothing formula has been proved:

Theorem 1.3.1 For $s \leq t$

$$\hat{\xi}_{s,t} = \hat{\xi}_{s,s} + \int_s^t I_{(\hat{\hat{h}}_u \neq 0)}(\hat{\hat{h}}_u)^{-1}(\Pi(\xi_s h_u) - \hat{\xi}_{s,u-}\hat{\hat{h}}_u)d\tilde{Q}_u$$

Prediction. The prediction problem is the derivation of a recursive expression for

$$\hat{\xi}_{t,s} = E[\xi_t | \underline{G}_s], \qquad t \geq s.$$

Again t is fixed and $0 \leq s \leq t \leq T$, so $\hat{\xi}_{t,s}$ is a square integrable $(\{\underline{G}_s\}, P)$ martingale.

Consequently, it has a representation

$$\hat{\xi}_{t,s} = \hat{\xi}_{t,o} + \int_o^s \gamma_{u,t} d\tilde{Q}_u \tag{1.3.25}$$

where $E[\int_o^t \gamma_{u,t}^2 \hat{\hat{h}}_u du] < \infty$ and $\{\gamma_{u,t}\}$ is $\{\underline{G}_u\}$ predictable.

Define $X_s = E[\xi_t | \underline{F}_s]$ and suppose the martingale X has no common jumps with N. Then $[X, N]_t = 0$.

Now

$$X_s N_s = \int_o^s N_{u-} dX_u + \int_o^s X_{u-} dN_u \tag{1.3.26}$$

Clearly, $\hat{\xi}_{t,s} = E[X_s | \underline{G}_s]$ so taking the $(\{\underline{G}_s\}, P)$ optional projection of (1.3.26)

$$\hat{\xi}_{t,s} N_s = \int_o^s E[X_{u-} h_u | \underline{G}_u] du + K_t^1, \tag{1.3.27}$$

where K_t^1 is a local $(\underline{\underline{G}}, P)$ martingale. Using (1.3.25) and the product rule

$$\begin{aligned}\hat{\xi}_{t,s} N_s &= \int_o^s \hat{\xi}_{t,u-} dN_u + \int_o^s N_{u-} \gamma_{u,t} d\tilde{Q}_u + \int_o^s \gamma_{u,t} dN_u \\ &= \int_o^s (\hat{\xi}_{t,u-} \hat{\hat{h}}_u + \gamma_{u,t} \hat{\hat{h}}_u) du + K_t^2 \end{aligned} \tag{1.3.28}$$

where K_t^2 is a local $(\underline{\underline{G}}, P)$ martingale. From (1.3.27) and (1.3.28), using the unique decomposition of special semimartingales, we have

$$\gamma_{u,t} = I_{(\hat{\hat{h}} \neq 0)}(\hat{\hat{h}}_u)^{-1}(E[\xi_t h_u | \underline{G}_u] - \hat{\xi}_{t,u-}\hat{\hat{h}}_u)$$

and we can state the following result.

Theorem 1.3.2 For $t \geq s$

$$\hat{\xi}_{t,s} = \hat{\xi}_{t,o} + \int_o^s I_{(\hat{\hat{h}}_u \neq 0)}(\hat{\hat{h}}_u)^{-1}(\Pi(\xi_t h_u) - \hat{\xi}_{t,u-}\hat{\hat{h}}_u)d\tilde{Q}_u.$$

1.4 THE ZAKAI FORMS OF THE PREDICTION AND SMOOTHING EQUATIONS

We suppose there are two probability measures P and P_1 related as in Section 1.2. The $(\underline{\underline{G}}, P)$ optional and predictable projections will be denoted by Π and Π_p, respectively; the $(\underline{\underline{G}}, P_1)$ optional and predictable projections will be denoted by $\hat{}$ and $\hat{\hat{}}$ respectively. Using this notation the smoothing formula of Theorem 1.3.1 states

$$\Pi(\xi_s)_t = E[\xi_s|\underline{G_t}] = \Pi(\xi_s)_s + \int_s^t I_{\Pi_p(h)\neq 0}(\Pi_p(h_u))^{-1}(\Pi_p(\xi_s h_u)_u - \Pi_p(\xi_s)_{u-}\Pi_p(h)_u)d\tilde{Q}_u. \tag{1.4.29}$$

Again, equation (1.4.29) is quadratic in the conditional distribution Π_p. Using the Baye's formula

$$\begin{aligned}\Pi(\xi_s)_t &= \frac{E_1[\xi_s\Lambda_t|\underline{G_t}]}{E_1[\Lambda_t|\underline{G_t}]}\\ &= \hat{\Lambda}_t^{-1}E_1[\xi_s\Lambda_t|\underline{G_t}].\end{aligned}$$

Write $\sigma_{s,t}(\xi_s) = E_1[\xi_s\Lambda_t|\underline{G_t}]$ for the unnormalized smoothed estimate of ξ_s. Then

$$\sigma_{s,t}(\xi_s) = \hat{\Lambda}_t\Pi(\xi_s)_t$$

and from (1.4.29) and Theorem 1.2.4

$$\begin{aligned}\sigma_{s,t}(\xi_s) &= \hat{\Lambda}_s\Pi(\xi_s)_s + \int_s^t \hat{\Lambda}_{u-}\gamma_{s,u}\, d\tilde{Q}_u\\ &\quad + \int_s^t \Pi(\xi_s)_{u-}\hat{\Lambda}_{u-}(\Pi_p(h_u) - 1)d\bar{Q}_u\\ &\quad + \int_s^t \hat{\Lambda}_{u-}\gamma_{s,u}(\Pi_p(h_u) - 1)dN_u\\ &= \sigma_{s,s}(\xi_s) + \int_s^t \hat{\Lambda}_{u-}(\Pi_p(\xi_s h_u)_u - \Pi_p(\xi_s)_{u-}\Pi_p(h)_u)d\bar{Q}_u\\ &\quad + \int_s^t \Pi(\xi_s)_{u-}\hat{\Lambda}_{u-}(\Pi_p(h)_u - 1)d\bar{Q}_u\\ &= \sigma_{s,s}(\xi_s) + \int_s^t \hat{\Lambda}_{u-}(\Pi(\xi_s h_u) - \Pi_p(\xi_s)_{u-})d\bar{Q}_u.\end{aligned}$$

Using Lemma 1.1, $\Pi(\xi_s h_u)$ can be replaced by $\Pi_p(\xi_s h_u)$ in (1.3.23) and consequent formulas, so we see that $\sigma_{s,t}$ satisfies the following linear Zakai equation, in which the $\Pi_p(h)^{-1}$ is also absent:

Theorem 1.4.1

$$\sigma_{s,t} = \sigma_{s,s}(\xi_s) + \int_s^t (\sigma_{s,u-}(\xi_s h_u) - \sigma_{s,u-}(\xi_{u-}))d\bar{Q}_u.$$

Prediction. Consider for $s \leq t$

$$\hat{\xi}_{t,s} = E[\xi_t | \underline{G}_s] = \frac{E_1[\Lambda_t \xi_t | \underline{G}_s]}{E_1[\Lambda_t | \underline{G}_s]}$$

and write $\sigma_{t,s}(\xi_t) = E_1[\Lambda_t \xi_t | \underline{G}_s]$ so $\sigma_{t,s}(\xi_t) = \hat{\Lambda}_s \hat{\xi}_{t,s}$.

From Theorems 1.2.4 and 1.3.2

$$\sigma_{t,s}(\xi_t) = \hat{\xi}_{t,o} + \int_o^s \hat{\Lambda}_{u-}\gamma_{u,t} d\tilde{Q}_u$$
$$+ \int_o^s \hat{\xi}_{t,u-}\hat{\Lambda}_{u-}(\Pi_p(h_u) - 1)d\bar{Q}_u + [\hat{\Lambda}, \hat{\xi}]_s$$

where

$$[\hat{\Lambda}, \hat{\xi}]_s = \sum_{0<u\leq s} \triangle\hat{\Lambda}_u \triangle\hat{\xi}_{s,u}$$
$$= \int_o^s \gamma_{u,t}\hat{\Lambda}_{u-}(\Pi_p(h_u) - 1)dN_u.$$

Therefore,

$$\sigma_{t,s}(\xi_t) = \sigma_{t,o}(\xi_t) + \int_o^s \hat{\Lambda}_{u-}(\Pi(\xi_t h_u) - \hat{\xi}_{t,u-}\Pi_p(h_u))d\bar{Q}_u$$
$$+ \int_o^s \hat{\xi}_{t,u-}\hat{\Lambda}_{u-}(\Pi_p(h_u) - 1)d\bar{Q}_u.$$

As above, this is

$$= \sigma_{t,o}(\xi_t) + \int_o^s \hat{\Lambda}_{u-}(\Pi_p(\xi_t h_u) - \hat{\xi}_{t,u-})d\bar{Q}_u.$$

So $\sigma_{t,s}(\xi_t)$ satisfies the following linear Zakai equation, in which $\Pi_p(h)^{-1}$ is absent:

Theorem 1.4.2

$$\sigma_{t,s}(\xi_t) = \sigma_{t,o}(\xi_t) + \int_o^s (\sigma_{t,u-}(\xi_t h_u) - \sigma_{t,u-}(\xi_t))d\bar{Q}_u.$$

1.5 CONTROL OF A PARTIALLY OBSERVED MARKOV CHAIN

As before, consider a finite state Markow process $\{X_t\}, 0 \leq t \leq T$, whose state space S is the set of unit vectors $\{e_1, \ldots, e_N\}$ of R^N. Write $p_t^i = P(X_t = e_i)$ and suppose there is a time dependent family of generators $A(t,u)$, which also depend on the control parameter $u \in U$, such that the probability column vector $p_t = (p_t^1,\dots,p_t^N)'$ satisfies the Kolmogorov forward equation

$$\frac{dp}{dt} = A(t,u)p_t.$$

Here U, the set of control values, is a compact metric space and A is required to satisfy suitable measurability conditions. N is a conditional Poisson point process with intersity $h(X_s)$, so as before

$$N_t = \int_o^t h(X_s)ds + Q_t = \int_o^t \hat{\hat{h}}(X_s)ds + \tilde{Q}_t,$$

where Q (resp. $\tilde{Q}$) is a $(P,\underline{\underline{F}})$ (resp. $(P,\underline{\underline{G}})$) martingale. The set $\underline{U}$ of admissible control functions $\{u\}$ is the set of $\underline{\underline{G}}$ predictable processes with values in U. This meas that, if $T_1,T_2,\dots$ are the jumps times of N, then for $T_n < t \leq T_{n+1}$ $u \in \underline{U}$ is function only of $T_1,T_2,\dots,T_n$ and t. For each control function $u \in \underline{U}$ the process M^u is a $(P,\underline{\underline{F}})$ martingale, where

$$M_t^u = X_t^u - X_o^u - \int_o^t A(s,u)X_s^u ds.$$

We assume that, almost surely, X^u, (and so M^u), has no jumps in common with N, so $[M^u,N]=0$.

If l is a real valued function on the state space S then l can be identified with the vector $(l^1,l^2,\dots,l^N)$, so that for $x \in R^N$ $l(x) = <l,x>$, where $<,>$ denotes the inner product in R^N. The control problem we wish to consider is that of choosing $u \in \underline{U}$ so that the expected cost

$$J(u) = E[<l,X_T^u>]$$

is minimized. (See the survey article by Davis [2]).

Write

$$\hat{p}_{s-}^i(u) = P(X_S^u = e_i|\underline{G}_s)$$

and recall that

$$\hat{\hat{h}}(X_s^u) = \sum_{i=1}^N h(e_i)\hat{p}_{s-}^i(u).$$

1.5.1 The separated form of the control problem

As in Example 1.1.9 we have the following equation for $\hat{p}_s = (\hat{p}_s^1, \dots, \hat{p}_s^N)'$:

$$\hat{p}_t = \hat{p}_o + \int_o^t A(s,u)\hat{p}_{s-}ds + \int_o^t I_{(\hat{\hat{h}}_s \neq 0)}(\hat{\hat{h}}_s)^{-1}(B\hat{p}_{s-} - \hat{\hat{h}}_s\hat{p}_{s-})\tilde{Q}_s \tag{1.5.30}$$

where B is the diagonal matrix with entries $h(e_i)$. Recall that

$$\hat{\Lambda}_t = 1 + \int_o^t \hat{\Lambda}_{s-}(\hat{\hat{h}}_s - 1)(dN_s - ds) \tag{1.5.31}$$

Again computing the product of (1.5.30) and (1.5.31) we have the linear Zakai equation for the unnormalized distribution

$$\begin{aligned} q_t &= E_1[\Lambda_t\, X_t \mid \underline{G}_t] \\ &= \hat{\Lambda}_t\, \hat{p}_t \\ &= \hat{p}_o + \int_o^t \hat{p}_{s-}.\hat{\Lambda}_{s-}(\hat{\hat{h}}_s - 1)(dN_s - ds) \\ &\quad + \int_o^t \hat{\Lambda}_{s-}\, A(s,u)\hat{p}_{s-}\, ds \\ &\quad + \int_o^t \hat{\Lambda}_{s-}\, I_{(\hat{\hat{h}} \neq 0)}(\hat{\hat{h}}_s)^{-1}(B\hat{p}_{s-} - \hat{\hat{h}}_s\hat{p}_{s-})(dN_s - \hat{\hat{h}}_s ds) \\ &\quad + \int_o^t \hat{\Lambda}_{s-}(\hat{\hat{h}}_s - 1)I_{\hat{\hat{h}} \neq 0}(\hat{\hat{h}}_s)^{-1}(B\hat{p}_{s-} - \hat{\hat{h}}_s\hat{p}_{s-})dN_s \\ &= \hat{p}_o + \int_o^t \hat{\Lambda}_{s-}\, \hat{p}_{s-}(\hat{\hat{h}}_s - 1)d\bar{Q}_s + \int_o^t \hat{\Lambda}_{s-}\, A(s,u)\hat{p}_{s-}\, ds \\ &\quad + \int_o^t \hat{\Lambda}_{s-}\, (B\hat{p}_{s-} - \hat{p}_{s-})d\bar{Q}_s \end{aligned}$$

Therefore

$$q_t = q_o + \int_o^t A(s,u)q_{s-}\, ds + \int_o^t (B-1)q_{s-}d\bar{Q}_s, \tag{1.5.32}$$

where, as in section 2, $Q_s = N_s - s.$ is $(P_1, \underline{\underline{G}})$ martingale.

Remarks 5.2 The q defined by (1.5.32) will depend on the control $u \in \underline{U}$; when necessary this will be denoted by writing q_t^u. Also, the exponential density Λ will depend on $u \in \underline{U}$, so we could write Λ^u. If a probability measure P_1^u is defined by

$$E[\frac{dP_1^u}{dP} | \underline{F}_t] = (\Lambda_t^u)^{-1}$$

then under P_1^u the process N is a standard Poisson process.

The expected cost if $u \in \underline{U}$ is used is

$$\begin{aligned} J(u) &= E[< l, X_T^u >] \\ &= E_1[\Lambda_T^u < l, X_T^u >] \\ &= E_1[< l, \Lambda_T^u \; X_T^u >] \\ &= E_1[< l, E_1[\Lambda_T^u X_T^u \mid \underline{G}_T] >] \\ &= E_1[< l, q_T^u >]. \end{aligned}$$

Here E_1, denotes expectation with respect to a probability measure P_1 under which N is a standard Poisson process. The control problem can, therefore, be expressed in the following separated form:

Minimize $J(u) = E_1[< l, q_T^u >]$ where q_t^u, $0 \le t \le T$, is given by the dynamics (1.5.32). Note that

$$q_o = p_o.$$

1.5.2 A minimum principle The problem has now been formulated in the form: choose $u \in \underline{U}$ to minimize $J(u) = E_1[< l, q_T^u >]$ where for $0 \le t \le T$

$$q_t^u = p_o + \int_o^t A(s,u) \, q_{s-}^u \, ds + \int_o^t (B-1) q_{s-}^u \, d\bar{Q}_s. \tag{1.5.33}$$

Under P_1 N is standard Poisson process and the equation (1.5.33) is in the 'stochastic open loop' form discussed by Bismut [1] and Kushner [7]. That is, the controls $u \in \underline{U}$ are adapted to the filtration generated by N (or $\bar{Q}$), and are not explicitly functions of the state q. Consequently, there are no problems concerning the existence of solutions to (1.5.33) for $u \in \underline{U}$. Indeed, the equation is linear in q.

For $u \in \underline{U}$ write $\Phi^u(t,s)$ for the fundamental matrix solution of

$$d\Phi^u(t,s) = A(t,s)\Phi^u(t,s)dt + (B-1)\Phi^u(t,s)(dN_t - dt) \tag{1.5.34}$$

with initial condition $\Phi^u(s,s) = I$, the $N \times N$ identity matrix. Then the solution of (1.5.33) can be expressed as

$$q_t^u = \Phi^u(t,o) q_o.$$

If $0 \le s \le t$ clearly, by uniqueness, $q_t^u = \Phi^u(t,s) q_s^u$. The expected cost can be written

$$J(u) = E_1[< l, \Phi^u(T,o) q_o >].$$

Suppose $u^* \in \underline{U}$ is an optimal control so that $J(u^*) \leq J(u)$, for all other $u \in \underline{U}$. For a fixed $u \in U$, s, ε, such that $0 \leq s < s+\varepsilon \leq T$, and $A \in \underline{G}_s$ define a **strong variation** $u \in \underline{U}$ of the optimal control $u^* \in \underline{U}$ by putting

$$u(t,\omega) = u^*(t,\omega) \text{ if } (t,\omega) \notin [s, s+\varepsilon] \times A$$

$$u(t,\omega) = u \text{ if } (t,\omega) \in [s, s+\varepsilon] \times A.$$

Now

$$J(u^*) = E_1[< l, \Phi^{u*}(T,o)q_o >]$$

and this can be written, with $'$ denoting transpose,

$$\begin{aligned} J(u^*) &= E_1[< l, \Phi^{u^*}(T, s+\varepsilon)\Phi^{u^*}(s+\varepsilon, s)\Phi^{u^*}(s,o)q_o >] \\ &= E_1[< \Phi^{u^*}(T, s+\varepsilon)' l, \Phi^{u^*}(s+\varepsilon, s)q_s^{u^*} >]. \end{aligned}$$

Similarly

$$J(u) = E_1[< \Phi^{u^*}(T, s+\varepsilon)' l, \Phi^{u}(s+\varepsilon, s)q_s^{u^*} >].$$

Now from (1.5.34)

$$\Phi^u(s+\varepsilon, s) = I + \int_s^{s+\varepsilon} A(r,u)\Phi^u(r,s)dr + \int_s^{s+\varepsilon} (B-1)\Phi^u(r,s)d\bar{Q}_r,$$

with a similar equation for $\Phi^{u^*}(s+\varepsilon, s)$.

Because $u^* \in \underline{U}$ is optimal

$$J(u) - J(u^*) \geq 0$$

So we have

$$\begin{aligned} &E_1[\langle \Phi^{u^*}(T, s+\varepsilon)' l, \int_s^{s+\varepsilon} (A(r,u)\Phi^u(r,s) - A(r,u^*)\Phi^{u^*}(r,s))ds\ q_s^{u^*} \rangle] \\ &+ E_1[\langle \Phi^{u^*}(T, s+\varepsilon)' l, \int_s^{s+\varepsilon} (B-1)(\Phi^u(r,s) - \Phi^{u^*}(r,s))d\bar{Q}_r\, q_s^{u^*} \rangle] \\ &\geq 0 \end{aligned} \tag{1.5.35}$$

for all s, ε and $A \in \underline{G}_s$.

Write K_1 for the first expectation above and K_2 for the second; also write

$$\Delta\Phi_r \text{ for } (B-1)(\Phi^u(r,s) - \Phi^{u^*}(r,s)).$$

The temptation is to argue that

$$E_1[\int_s^{s+\varepsilon} \triangle\Phi_r \, d\bar{Q}_r|\underline{G}_s] = 0; \tag{1.5.36}$$

however, this cannot be done immediately. Instead, write $\Gamma = \int_s^{s+\varepsilon} \triangle\Phi_r \, d\bar{Q}_r \, q_s^{u^*}$ so

$$K_2 =$$

$$\begin{aligned}
&E_1[\langle\Phi^{u^*}(T, s+\varepsilon)'l, \Gamma\rangle] \\
&= E_1[\langle\Phi^{u^*}(s+\varepsilon, s)'^{-1}\Phi^{u^*}(T,s)'l, \Gamma\rangle] \\
&\quad - E_1[\langle\Phi^{u^*}(s+\varepsilon, s)'^{-1}E_1[\Phi^{u^*}(T,s)'l|\underline{G}_s], \Gamma\rangle] \\
&\quad + E_1[\langle\Phi^{u^*}(s+\varepsilon, s)'^{-1}E_1[\Phi^{u^*}(T,s)'l|\underline{G}_s], \Gamma\rangle] \\
&\quad - E_1[\langle E_1[\Phi^{u^*}(s+\varepsilon, s)'^{-1}\Phi^{u^*}(T,s)'l|\underline{G}_s], \Gamma\rangle] \\
&\quad + E_1[\langle E_1[\Phi^{u^*}(s+\varepsilon, s)'^{-1}\Phi^{u^*}(T,s)'l|\underline{G}_s], \Gamma\rangle].
\end{aligned} \tag{1.5.37}$$

If M_r is the martingale $E_1[\Phi^{u^*}(T,s)'l|\underline{G}_r]$ for $s \le r \le T$ then $M_{s+\varepsilon}$ has a representation as a stochastic integral $M_s + \int_s^{s+\varepsilon} \gamma_r d\bar{Q}_r$ for some predictable vector integrand γ. Also

$$d\Phi^{u^*}(r,s)^{-1} = -\Phi^{u^*}(r,s)^{-1}A(r,u)dr - \Phi^{u^*}(r,s)^{-1}(B-I)d\bar{Q}_r$$

and

$$\Phi^{u^*}(r,s) = (I + (B-I)\Delta N_r)\Phi^{u^*}(r,s),$$

so

$$\begin{aligned}
\Phi^{u^*}(s+\varepsilon, s)'^{-1} =& I - \int_s^{s+\varepsilon} \\
&A(r,u)'\Phi^{u^*}(r-,s)'^{-1}dr + (B^{-1}-I)\int_s^{s+\varepsilon} \Phi^{u^*}(r-,s)'^{-1}d\bar{Q}_r \\
&-(B-I)(B^{-1}-I)\int_s^{s+\varepsilon} \Phi^{u^*}(r-,s)'^{-1}dr.
\end{aligned}$$

Because $E_1[\Gamma|\underline{G}_s] = 0$ the final term in (1.5.37) is zero. The remaining terms in (1.5.37) give rise to inner products of the form

$$E_1[\langle\int_s^{s+\varepsilon} \alpha_r dr, \Gamma\rangle]$$

and

$$E_1[\langle \int_s^{s+\varepsilon} \beta_r d\bar{Q}_r, \Gamma \rangle],$$

for some integrands α and β. Now write $\Gamma_t = \int_s^t \Delta\Phi_r d\bar{Q}_r q_s^{u^*}$ so $\Gamma = \Gamma_{s+\varepsilon}$. Although we are dealing with vector integrals, the inner product of Γ and these stochastic integrals gives rise to integrals with respect to $d\bar{Q}$, which are martingales, and integrals of the form

$$\int_s^{s+\varepsilon} \alpha_r \Gamma_{r-}\, dr$$

and

$$\int_s^{s+\varepsilon} \beta_r \Delta\Phi_r dr\, q_s^{u^*}.$$

Taking the E_1 expectation the martingale term gives zero; dividing by $\varepsilon > 0$ and letting $\varepsilon \to 0$ the dr integrals give terms of the form

$$E_1[\alpha_s \Gamma_{s-}] \text{ and } E_1[\beta_s \Delta\Phi_s q_s^{u^*}].$$

However, when $r = s$, $\Gamma_{s-} = \Delta\Phi_s = 0$. The limit obtained by dividing (1.5.35) by ε and letting $\varepsilon \to 0$ is, therefore,

$$E_1[\langle \Phi^{u^*}(T,s)'l, (A(s,u(s,w)) - A(s,u^*(s,w))q_s^{u^*} \rangle]$$
$$\text{and this is} \geq 0. \quad (1.5.38)$$

However, $u(s,w) = u^*(s,w)$ if $w \notin A \in \underline{G}_s$. Write

$$P_s^{u^*} = E_1[\Phi^{u^*}(T,s)'l | \underline{G}_s]. \qquad (1.5.39)$$

Then (1.5.38) can be written

$$E_1[< P_s^{u^*}, (A(s,u) - A(s,u^*))q_s^{u^*} > I_A] \geq 0$$

for all $u \in U$, $s \in [0,T]$, $\varepsilon > 0$ and $A \in \underline{G}_s$. That is, the optimal control $u^* \in \underline{U}$ satisfies the following minimum principle:

$$\langle A(s,u^*)' P_s^{u^*},\ q_s^{u^*} \rangle = min_{u \in U} \langle A(s,u)' P_s^{u^*},\ q_s^{u^*} \rangle.$$

Here $P_s^{u^*}$ plays the role of the co-state variable; from its definition (5.8) it is given by the variation in the cost criterion in the sense that

$$\nabla\langle l, X\rangle = l.$$

1.6 AN APPROXIMATE MINIMUM PRINCIPLE

The minimum principle above requires the existence of an optimal control $u^* \in \underline{U}$. Using a variational principle of Ekeland [3] an approximate minimum principle will now be obtained. Write

$$J = inf_{u \in \underline{U}} \ J(u).$$

Then for any $\delta > 0$ there is always a control u such that

$$J \leq J(u) \leq J + \delta.$$

For any to admissible controls $u_1, u_2, \in \underline{U}$ write

$$C = \{(s,w) \in [o,T] \times \Omega : u_1(s,w) \neq u_2(s,w)\}$$

and $d(u_1,u_2)$ for the Lebesgue $\times P_1$ measure of C.

Then, as in Elliott and Kohlmann [5], $\underline{U}$ is a complete metric space under d. (See also Kohlmann [6].) Furthermore if, for example, A is bounded, $J(u)$ is a continuous function on $(\underline{U}, d)$. (See Lemma 6.1 below). The variational principle of Ekeland [3] states that for any $\delta > 0$ there is a $v^\delta \in \underline{U}$ such that

$$J \leq J(v^\delta) \leq J + \delta$$

and for any other control $u \in \underline{U}$

$$J(u) \geq J(v^\delta) - \delta \ d(v^\delta, u).$$

Write Φ^δ for Φ^{v^δ} and q^δ for $q^{v^\delta} = \Phi^\delta(t,o)q_o$. Again, consider a fixed value $u \in U$, $s \in [o,T)$, $\varepsilon > 0$ and $A \in \underline{G}_s$, and define a strong variation $u \in \underline{U}$ of v^δ by putting

$$u(t,w) = v^\delta(t,w) \text{ if } (t,w) \notin [s, s+\varepsilon] \times A$$

$$u(t,w) = u \text{ if} (t,w) \in [s, s+\varepsilon] \times A. \tag{1.6.40}$$

Then $d(v^\delta, u) \leq \varepsilon P_1(A) \leq \varepsilon$ so

$$J(u) - J(v^\delta) \geq -\delta\varepsilon. \tag{1.6.41}$$

As before,

$$J(v^\delta) = E_1[\langle \Phi^\delta(T, s+\varepsilon)'l, \Phi^\delta(s+\varepsilon, s)q_s^\delta\rangle]$$

and

$$J(u) = E_1[\langle \Phi^\delta(T, s+\varepsilon)'l, \Phi^u(s+\varepsilon, s)q_s^\delta\rangle].$$

Write $p_s^\delta = E_1[\Phi^\delta(T, s)'l \mid \underline{G}_s]$. Substituting in (1.6.40), dividing by ε and letting $\varepsilon \to 0$ we obtain as before that the δ- optional control v^δ satisfies

$$\langle A(s,u)'p_s^\delta, q_s^\delta\rangle - \langle A(s, v^\delta)'p_s^\delta, q_s^\delta\rangle \geq -\delta$$

$$\text{for all } u \in U.$$

That is

$$\langle A(s,v^\delta)'p_s^\delta, q_s^\delta\rangle \ \leq min_{u\in U}\langle A(s,u)'p_s^\delta, q_s^\delta\rangle + \delta$$

so v^δ minimizes to within δ the Hamiltonian at each time $s \in [o, T]$.

Lemma 1.6.1 Suppose the set of matrices $\{A(s,u)\}$ is bounded. Consider the set $\underline{U}$ of admissible controls with metric d. Then $J : \underline{U} \to R$ is a continuous function on $(\underline{U}, d)$.

Proof: Suppose $\{u_n\}$ is a sequence in $(\underline{U}, d)$ which converges to $u^* \in \underline{U}$. Write q^* for q^{u^*}. Then from (1.5.33)

$$q_T^* = q_o + \int_o^T A(s,u^*)q_{s-}^* ds + \int_o^T (B-1)q_{s-}^* d\bar{Q}_s,$$

and

$$q_T^{u_n} = q_o + \int_o^T A(s,u_n)q_{s-}^{u_n} ds + \int_o^T (B-1)q_{s-}^{u_n} d\bar{Q}_s.$$

Write $q_s^* - q_s^{u_n} = \Delta q_n(s)$ and $\|X\|^2 = X_1^2 + \ldots + X_N^2$ for $X \in R^N$.

Then

$$\begin{aligned}
\Delta q_n(T) &= \int_o^T (A(s,u^*)q_{s-}^* - A(s,u_n)q_{s-}^{u_n})ds \\
&\quad + \int_o^T (B-1)\Delta q_n(s-)d\bar{Q}_s \\
&= \int_o^T (A(s,u^*) - A(s,u_n))q_s^* ds + \int_o^T A(s,u_n)\Delta q_n(s-)ds \\
&\quad + \int_o^T (B-1)\Delta q_n(s-)d\bar{Q}_s,
\end{aligned}$$

so for some constant C_1

$$\|\triangle q_n(T)\|^2 \leq C_1(\|\int_o^T (A(s,u^*) - A(s,u_n))q^*_{s-}ds\|^2$$

$$+ \|\int_o^T A(s,u_n)\triangle q_n(s-)ds\|^2 + \|(B-1)\int_o^T \triangle q_n(s-)d\bar{Q}_s\|^2).$$

Therefore, taking expectations under P_1,

$$E_1[\|\triangle q_n(T)\|^2] \leq C_2(\int_o^T E_1(\|A(s,u^*) - A(s,u_n)\|^2 \ \|\triangle q_n(s-)\|^2)ds$$

$$+ \int_o^T E_1(\|A(s,u_n)\|^2 \ \|\triangle q_n(s-)\|^2 ds$$

$$+ \|B-1\|^2 \int_o^T E_1(\|\triangle q_n(s-)\|^2)ds)$$

Write $B_n = \int_o^T E_1(\|A(s,u^*) - A(s,u_n)\|^2 \ \|\triangle q_n(s-)\|^2)ds$;
then $B_n \leq C_3 d(u_n, u^*)$, because A is bounded. By Gronwall's inequality

$$E_1(\|\triangle q_n(T)\|^2) \leq C_4 B_n \ \exp \ C_5 T.$$

Therefore, $lim_{n\to\infty}\|\triangle q_n(T)\|^2 = 0$.

Now $J(u^*) - J(u_n) = E_1(< l, \triangle q_n(T) >)$

so $|J(u^*) - J(u_n)| \leq C_6 \ E_1(\|\triangle q_n(T)\|^2)$

and $lim_{n\to\infty} J(u_n) = J(u^*)$.

REFERENCES

[1] BISMUT, J. M., (1978). *An introductory approach to duality in optimal stochastic control.* SIAM Review 20, 62-78.

[2] DAVIS, M. H. A., *Some current ideas in stochastic control theory.* Stochastic.

[3] EKELAND, I. (1979). *Non convex minimization problems.* Bull AMS 1, 443-474.

[4] ELLIOTT, R.J., (1982). *Stochastic Calculus and Applications.* Springer-Verlag. New York, Heidelberg, Berlin.

[5] ELLIOTT, R. J. and KOHLMANN, M., (1980). *The variational principle and stochochastic optimal control.* 3, 229-241.

[6] KOHLMANN, M., (1985). *A maximum principle and dynamic programming for a partially observed control problem.* Discussion paper. Universität Konstanz.

[7] KUSHNER, H. J., (1972). *Necessary conditions for continuous parameter stochastic optimization problems.* SIAM. J. Control 10, 550-565.

[8] LOÈVE, M., *Probability Theory*, Vols. I and II, 4th Ed. Springer-Verlag, Berlin, Heildelberg, New York.

2

NONLINEAR FILTERING THEORY

Alain Bensoussan

INTRODUCTION

We present in these lectures the theory on non linear filtering. Given a signal governed by a diffusion

$$dx_t = g(x_t, t)dt + \sigma(x_t, t)dw_t \qquad x(0) = \xi$$

and an observation process described by

$$dz_t = h(x_t, t)dt + db_t \qquad z(0) = 0$$

the problem at stake is to compute the conditional probability

$$\pi(t)(\Psi) = \mathrm{E}[\Psi(x_t)|z(s), s \leq t]$$

where Ψ is any test function.

The main core of the theory consists in characterizing $\pi(t)(\Psi)$ as the solution of an evolution equation. We study this equation and prove existence and uniqueness in some adequate space of measures. Moreover, the conditional probability can be written as a ratio

$$\pi(t)(\Psi) = \frac{p(t)(\Psi)}{p(t)(1)}$$

where $p(t)(\Psi)$ is called the unnormalized conditional probability. It is also characterized as the solution of an evolution equation, which is in principle simpler than the other one, since it is linear.

These equations are known as Kushner-Stratonovitch and Zakai equations. We treat the cases when the signal noise and the observation noises are correlated or not.

When there is a density, we can write

$$p(t)(\Psi) = \int \Psi(x)p(x,t)dx$$

and the process $p(x,t)$ will be the solution of a stochastic P.D.E. In fact, when there is no correlation and h is sufficiently smooth, it is possible by a transformation called the robust form, to reduce this equation to an ordinary P.D.E. with random coefficients. We develop this approach and use it to give representation formulas for $p(x,t)$. These formulas can be viewed as the analogue of the probabilistic interpretation of the solution of parabolic equations.

The stochastic P.D.E. can also be studied directly, without using the robust form. It is possible to introduce forward and backward stochastic P.D.E. which are linked by a duality relation, extending the classical situation with parabolic P.D.E.

2.1 NON LINEAR FILTERING EQUATIONS

2.1.1 Setting of the problem—notation

Let Ω, A, P, F^t be a usual system, where F^t is an increasing family of sub σ-algebras of A, continuous to the right and complete for the probability P. We assume that there exist $w(\cdot), b(.)$ two independent Wiener processes with values in R^n, R^m, whose covariance is respectively $Q(.), R(.)$.

Let also ξ be a random variable with values in R^n, which is F^o measurable, independent of $w(.), b(.)$. Its probability measure is denoted by π_o.

Let also $g(x,t), \sigma(x,t)$, $h(x,t)$ be functions such that

$$|g(x_1,t) - g(x_2)| \le k|x_1 - x_2| \tag{2.1.1}$$

$$||\sigma(x_1,t) - \sigma(x_2)|| \le k|x_1 - x_2|$$

$$g(0,t), \sigma(0,t) \text{ are bounded}$$

$$g, h \text{ Borel functions}$$

$$h(x,t) \text{ is Borel and } |h(x,t)| \le k(1+|x|). \tag{2.1.2}$$

We also assume that

$$\mathrm{E}|\xi|^2 = \int |x|^2\pi_o(dx) < \infty. \tag{2.1.3}$$

Let us define two processes x_t, z_t defined by the following relations

$$dx_t = g(x_t, t)dt + \sigma(x_t, t)dw_t$$

$$x(0) = \xi \tag{2.1.4}$$

$$dz = h(x_t, t)dt + db_t$$

$$z(0) = 0 \tag{2.1.5}$$

The process x_t is uniquely defined by virtue of the assumptions (2.1.1). It is called the *signal process.* The z_t is called the *observation process.* We define

$$Z^t = \sigma(z(s),\ s \leq t)$$

which is the *observation filtration.* We have $Z^t \subset F^t$. Let Ψ be a Borel bounded function on R^n, the problem of non linear filtering consists in finding

$$\pi(t)(\Psi) = E[\Psi(x_t)|Z^t]. \tag{2.1.6}$$

The map $\pi(t)$ is a random measure on R^n for any t. We thus have defined a stochastic process with values in the space of positive measures on R^n. Clearly, one has

$$\pi(0)(\Psi) = \pi_0(\rho) = \int \Psi(x)\pi_o(dx). \tag{2.1.7}$$

Change of probability measure. We first perform a change of probability measure on Ω, A, which transforms $z(.)$ into a Wiener process. For that purpose we introduce the stochastic process

$$\rho_t = \exp\left\{\int_o^t -h^*(x_s)R_s^{-1}db_s - \frac{1}{2}\int_o^t h^*(x_s)R_s^{-1}h(x_s)ds\right\}. \tag{2.1.8}$$

We first state the

Lemma 2.1.1 One has

$$E\rho_t = 1. \tag{2.1.9}$$

Sketch of the proof. The process ρ_t satisfies the relation

$$d\rho_t = -\rho_t h^*(x_t, t)R_t^{-1}db_t.$$

The only difficulty is due to the fact that h is not bounded. Indeed if h were bounded, then in fact $E\rho_t^2 < \infty$, and thus $E\rho_t = 1$, because $\rho_o = 1$ and since we can take the expectation of the stochastic integral, which is 0. For non bounded h, we compute $d\frac{\rho_t}{1+\varepsilon\rho_t}$, hence

$$E\frac{\rho_t}{1+\varepsilon\rho_t} = 1 - E\int_o^t \frac{\varepsilon\rho_s^2 h^*(x_s)R_s h(x_s)ds}{(1+\varepsilon\rho_s)^3}. \tag{2.1.10}$$

But we may prove independently that $E\rho_t|x_t|^2 < C$. We then use Lebesgue's theorem to assert that the right hand side of (2.1.10) tends to 1. The property (2.1.9) follows immediately.

From (2.1.9) we can then define a probability $\tilde{P}$ on (Ω, A) such that

$$\frac{d\tilde{P}}{dP}|_{F^t} = \rho_t. \tag{2.1.11}$$

For the system $(\Omega, A, \tilde{P}, F^t)$ the processes $w(.)$, $z(.)$ become F^t Wiener processes, with covariance $Q(.)$, $R(.)$, independent of ξ.

Define

$$\eta_t = \frac{1}{\rho_t} = \exp\left\{\int_0^t -h^*(x_s)R_s^{-1}dz_s - \frac{1}{2}\int_o^t h^*(x_s)R_s^{-1}h(x_s)ds\right\}$$

and naturally

$$\frac{dP}{d\tilde{P}}|_{F^t} = \eta_t.$$

It is then possible to express $\pi(t)(\Psi)$ by the following formula, called the *Kallianpur-Striebel* formula

$$\begin{aligned}\pi(t)(\Psi) &= \frac{\tilde{E}(\Psi(x_t)\eta_t|Z^t)}{\tilde{E}(\eta_t|Z^t)} \\ &= \frac{p(t)(\Psi)}{p(t)(1)}\end{aligned} \tag{2.1.12}$$

defining $p(t)$ by

$$p(t)(\Psi) = \tilde{E}(\Psi(x_t)\eta_t|Z^t) \tag{2.1.13}$$

called the *unnormalized conditional probability.* We have defined a new process $p(t)$ with values in the space of positive measures on R^n. It is important, to avoid technical difficulties, to make precise the dependence of p with respect to the pair t, ω. To that extent, it is useful to define the following functional spaces

$$L_Z^r(0, T) = \left\{\theta(t; \omega) \in L^r((0, T) \times \Omega; dt \otimes \tilde{d}P) | \text{a.e. } t, \theta_t \in L^r\left(\Omega, Z_t, \tilde{P}\right)\right\}.$$

We then have the

Lemma 2.1.2 Let $\Psi(x, t)$ be a Borel function such that

$$|\Psi(x, t)| \leq K(1 + |x|^2).$$

Then

$$p(t)(\Psi(t)) = \tilde{E}(\Psi(x_t, t)\eta_t | Z^t)$$

can be considered as the representation of an element of $L_Z^1(0, T)$. If h is bounded, it is the representative of an element of $L_Z^2(0, T)$.

2.1.3 Zakai equation

Let us define the family of 2nd order differential operators

$$A(s) = g^* D - \frac{1}{2} tr\sigma Q\sigma * D^2 \tag{2.1.15}$$

then we can state the following

Theorem 2.1.1 We make the assumptions (2.1.1), (2.1.2), (2.1.3). Let $\Psi(x, t)$ belong to $C_b^{2.1}(R^n \times (0, \infty))$, then one has

$$\begin{aligned} p(t)(\Psi(t)) &= \pi_o(\Psi(0)) + \int_o^t p(s)(\frac{\partial \Psi}{\partial s} - A(s)\Psi)ds \\ &+ \int_o^t p(s)(h^*(s)\Psi(s))R_s^{-1} dz, \quad \forall t, \text{ a.s.} \end{aligned} \tag{2.1.16}$$

There is a technical difficulty in the formulation (2.1.16). The integrand $p(s)(h(s)\Psi(s))$ belonging only to $L_Z^1(0, T; R^n)$, what is the *meaning* of the stochastic integral?

Note that when h is bounded, then the integrand belongs to $L_Z^2(0, T;\ R^n)$ and the stochastic integral is defined as usual.

In the proof of Theorem 2.1.1 and more generally in the sequel, one makes a constant use of the following result. First, define for any $\beta(.) \in L^{\infty}(0,\, T;\, R^m)$ (not random) the martingale with values in the space of complex variables

$$\theta_t = \exp\left\{ i \int_o^t \beta^* R^{-1} dz + \frac{1}{2}\int_o^t \beta^* R^{-1} \beta ds \right\}. \tag{2.1.17}$$

We shall consider $L^1(\Omega,\, Z^T; \tilde{P})$, extended to complex valued variables. We then have the

Lemma 1.3 Let $\xi_T \in L^1(\Omega,\, Z^T,\, \tilde{P})$. Suppose that

$$E\xi_T\theta_T = 0, \quad \forall \beta(.) \tag{2.1.18}$$

then necessarily $\xi_T = 0$.

Sketch of the proof of Theorem 2.1.1

We give it only in the case when h, σ are bounded, avoiding the technical difficulty mentioned above. By Ito's formula, we have

$$d\,\theta_t\eta_t\Psi(x_t) = [\theta_t\eta_t(\frac{\partial\Psi}{\partial t} - A\Psi) + i\,\theta\Psi\eta h^* R^{-1}\beta]dt$$
$$+\theta_t\eta_t D\Psi^*\sigma dw - \theta_t\Psi\eta_t h^* R^{-1} dz$$

hence

$$\tilde{E} \quad \theta_t\eta_t\Psi(x_t) = \pi_o(\Psi(0)) + \tilde{E}\left[\theta\eta\left(\frac{\partial\Psi}{\partial s} - A(s)\Psi\right) + i\theta\Psi\eta h^* R^{-1}\beta\right] ds. \tag{2.1.19}$$

But

$$\tilde{E} \quad \theta_t\eta_t\Psi(x_t) = \tilde{E}\theta(t)p(t)(\Psi(t))$$
$$\tilde{E}\int_o^t \theta\eta\left(\frac{\partial\Psi}{\partial s} - A(s)\Psi\right) ds = \tilde{E}\int_o^t \theta_s p(s)\left(\frac{\partial\Psi}{\partial s} - A(s)\Psi\right) ds$$
$$= \tilde{E}\theta_t \int_o^t p(s)\left(\frac{\partial\Psi}{\partial s} - A(s)\Psi\right) ds$$

and

$$\tilde{E}\int_o^t i\theta\,\eta\Psi\,h^*R^{-1}\beta ds = \tilde{E}\int_o^t i\theta_s p(s)(\Psi h^*)R^{-1}\beta\,ds$$
$$= \tilde{E}\theta_t\int_o^t p(s)(\Psi(s)h^*(s))R_s^{-1}dz.$$

Collecting these results in (2.1.19), and making use of Lemma 2.1.3, the desired result follows.

2.1.4 Kushner - Stratonovitch equation

The K.S. equation is the analogue of Zakai equation for the measure $\pi(t)$. We state it as follows

Theorem 2.1.2 We make the assumptions of Theorem 2.1.1. Let $(\Psi(x,t) \in C_b^{2.1}(R^n \times (0,\infty))$, then one has

$$\pi(t)(\Psi(t)) = \pi_o(\Psi(0)) + \int_o^t \pi(s)\left(\frac{\partial\Psi}{\partial s}\right) - A(s)\Psi(s)ds +$$

$$\int_o^t [\pi(s)(h^*(s)\Psi(s)) - \pi(s)(\Psi(s))\pi(s)(h^*(s))]\,R_s^{-1}(dz - \pi(s)(h(s))ds), \quad \forall t, \text{a.s.} \tag{2.1.20}$$

There is an important remark to make in comparing (2.1.20) to (2.1.16). There is no difficulty in the definition of the stochastic integrals entering in the right hand side of (2.1.20). Indeed, one has the estimate

$$E|\pi(t)(h(t)\Psi(t))|^2 \leq C(1 + E|x_t|^2) \tag{2.1.21}$$

and thus $\int_o^t \pi(s)(h^*(s)\Psi(s))R_s^{-1}db$ is well defined. The other stochastic integrals are also well defined.

Sketch of the proof of Theorem 2.1.2. We give it in the case h, σ bounded. It consists in using Zakai equation together with the definition

$$\pi(t)(\Psi(t)) = \frac{p(t)(\Psi(t))}{p(t)(1)}.$$

The result then follows from Ito's calculus.

2.1.5 Innovation approach

The K. S. equation has been derived from Zakai equation. It is no more a linear equation and thus more complex than Zakai equation. On the other hand, at least when h is unbounded, the K. S. equation does not contain any difficulty that exists in the Z. equation as far as the definition of stochastic integrals is concerned. Furthermore, it seems useful to avoid the change of probability, since the K. S. equation gets rid of it.

These considerations motivate the following question. Is there a direct approach to the K. S. equation? The answer is yes, and in fact the K. S. equation was known before the Z. equation. The direct derivation comes from the *innovation approach*.

We begin by defining the *innovation process*, by the formula

$$v_t = z_t - \int_o^t \pi(s)(h(s))ds. \tag{2.1.22}$$

We have the

Lemma 2.1.4 The process ν_t is a P, Z^t Wiener process, with covariance $R(.)$.

Sketch of the proof. From the definition, one has

$$d\nu_t = db_t + (h(x_t) - \pi(t)(h))dt.$$

Let ϕ belong to $L^2(0, T; R^m)$ (not random), and set

$$\zeta_t = \exp i \int_o^t \phi_s^* d\nu_s.$$

By using Ito's calculus, we can check that

$$E\left[\exp i \int_o^t \phi_\lambda^* d\nu_\lambda | Z^t\right] = \exp - \int_t^s \phi_\lambda^* R_\lambda \phi_\lambda d\lambda$$

and this suffices to imply the desired result. The next step consists in making use of a representation formula, due to Fujisaki-Kallianpur-Kunita [1], cf. also Liptser-Shiryaev [5]. Let us define

$$N^t = \sigma(\nu_s,\, s \leq t)$$

which is included in Z^t. But $N^t \neq Z^t$ in general. However, there is the following special feature

$$dz = \pi_t(h)dt + d\nu_t \tag{2.1.23}$$

and $\pi_t(h)$ is adapted to Z^t, satisfying $E|\pi_t(h)|^2 < C$. Since ν_t is a Wiener process, we can refer to the Kunita-Watanabe [4] representation theorem, to assert that any square integrable P, Z^t martingale μ_t is a stochastic integral, i.e., can be written as

$$\mu_t = \mu_o + \int_o^t \lambda_s^* R_s^{-1} d\nu_s, \text{ with } \lambda_t \in L_N^2(0, T; R^m), \quad \forall T. \tag{2.1.24}$$

Because of the special form (2.1.23), the representation formula (2.1.24) extends to square integrable P, Z^t martingales, with an integrand $\lambda_t \in L_Z^2(0, T; R^m)$. We can then proceed with the

Innovation approach to the proof of Theorem 2.1.2. Consider the process

$$M_t = \pi(t)(\Psi(t)) - \int_o^t \pi(s)\left(\frac{\partial \Psi}{\partial s} - A(s)\Psi\right) ds.$$

It is a bounded process adapted to Z^t. Moreover, it is a P, Z^t martingale. Therefore, from the above representation formula (2.1.24) we can write

$$\begin{aligned} \pi(t)(\Psi(t)) = \pi_o(\Psi(o)) &+ \int_o^t \pi(s)(\frac{\partial \Psi}{\partial s} - A(s)\Psi)ds \\ &+ \int_0^t \lambda_s^* R_s^{-1} \; (dz_s - \pi(s)(h(s))ds) \end{aligned} \tag{2.1.25}$$

and λ_t is Z^t adapted, with $E\int_o^t |\lambda_t|^2 dt < \infty$. It remains to identify λ_t. Consider the process $\xi_t = \int_o^t \gamma_s^* R_s^{-1} d\nu_s$, where γ_t is Z^t adapted and bounded. Then one has

$$EM_t\xi_t = E\int_o^t \gamma_s^* R_s^{-1} \lambda_s ds.$$

However, a direct computation is possible and yields

$$EM_t\xi_t = E\int_o^t \gamma_s^* R_s^{-1} \left[\pi(s)(h(x_s))(\Psi(s)) - \pi(s)(h(x_s))\pi(s)(\Psi(x_s))\right] ds$$

and by comparison, one identifies λ_t.

Remark 2.1.1 This approach is also possible for Zakai euqation at least when h is bounded. It does not avoid the technical difficulties pointed out before.

Remark 2.1.2 The innovation approach seems inadequate for proving uniqueness.

2.2 UNBOUNDED COEFFICIENTS

We develop in this section the proofs of Theorems 2.1.1 and 2.1.2, in the general case, i.e., without assuming that σ, h are bounded.

2.2.1 Proof of Theorem 2.1.1

We proceed in several steps.

a. approximation

Let us define

$$\eta^\varepsilon = \frac{\eta}{1+\varepsilon\eta}.$$

Reasoning as in the case of bounded coefficients, we can prove that

$$\tilde{E}^{Z^t}(\Psi(x_t)\eta_t(1+\varepsilon\eta_t)^{-1} =$$

$$\frac{\pi_o(\Psi(0))}{1+\varepsilon} + \int_o^t \tilde{E}^{Z_s}\left[\eta(1+\varepsilon\eta)^{-1}\left(\frac{\partial\Psi}{\partial s} - A(s)\Psi\right) - 2\varepsilon\Psi\eta 2(1+\varepsilon\eta)^{-3}h^*R^{-1}h\right] ds$$

$$+ \int_o^t \tilde{E}^{Z^s}\left[\Psi\eta(1+\varepsilon\eta)^{-2}h^*\right] R^{-1}dz, \ \forall t, \text{ a.s.}$$

We can then derive the following limits

$$\tilde{E}^{Z^t}(\Psi(x_t)\eta_t(1+\varepsilon\eta_t)^{-1} \to p(t)(\Psi(t)) \text{ a.s.}$$

$$\int_o^t \tilde{E}^{Z_s}\left[\eta(1+\varepsilon\eta)^{-1}\left(\frac{\partial\Psi}{\partial s} - A(s)\Psi\right)\right] ds \to$$

$$\int_o^t p(s)\left(\frac{\partial\Psi}{\partial s} - A(s)\Psi\right) ds, \text{ a.s.}$$

$$\varepsilon\int_o^t \tilde{E}^{Z_s}\left[\Psi\eta^2(1+\varepsilon\eta)^{-3}h^*R^{-1}h\right] ds \to 0, \text{ a.s.}$$

There is a difficulty in passing to the limit in the stochastic integral, since we can guarantee only that

$$\tilde{E}^{Z^s}\left[\Psi\eta(1+\varepsilon\eta)^{-2}h\right] \to \tilde{E}^{Z_s}\left[\Psi\eta h\right]$$

$$\text{a.e., a.s. and in } L^1_Z(0, T; R^m). \tag{2.2.2}$$

b. a priori bound

We know that $\tilde{E}^{Z^s}[\Psi\eta h]$ belongs to $L^1_Z(0, T; R^m)$, which does not suffice to give a meaning to the stochastic integral $\int_o^t p(s)(\Psi(s)h^*(s))R_s^{-1}dz$, where of course from the definition of $p(s)$ (see (2.1.13)).

$$p(s)(\Psi(s)h^*(s)) = \tilde{E}^{Z^s}\left[\Psi(x_s)h^*(x_s)\right].$$

We shall prove in fact that

$$\text{a.s. } \tilde{E}^{Z^s}\left[\Psi(x_s)\eta_s h(x_s)\right] \text{ belongs to } L^\infty(0, T; R^m). \tag{2.2.3}$$

It is sufficient to prove that

$$\text{a.s. } \tilde{E}^{Z^S}\left(\eta_s(1+(x_s)^2)^{\frac{1}{2}}\right) \in L^\infty(0, T). \tag{2.2.4}$$

We apply (2.2.1) to $\Psi = \chi = (1+|x|^2)^{\frac{1}{2}}$. In fact, a slight generalization of (2.2.1) is needed since χ is unbounded. We deduce that the process

$$M_t^\varepsilon = \tilde{E}^{Z^t}\left[\chi(x_t)\eta_t(1+\varepsilon\eta_t)^{-1}\right] - \frac{\pi_o(\chi)}{1+\varepsilon}$$

$$-\int_o^t \tilde{E}^{Z^t}\left[-\eta(1+\varepsilon\eta)^{-1}A_\chi - 2\varepsilon\chi\eta^2(1+\varepsilon\eta)^{-3}h^*R^{-1}h\right]d\lambda$$

is a $Z^t\tilde{P}$ martingale.

One can then deduce that the process

$$M_t = \tilde{E}^{Z^t}(\chi(x_t)\eta_t) - \pi_o(\chi) + \int_o^t \tilde{E}^{Z^S}(\eta A\chi)ds$$

is a $\tilde{P}$, Z^t martingale, and $\tilde{E}M_t = 0$.

From a result of Meyer [6], it follows that M_t has a modification which is continuous to the right. Moreover, a right continuous martingale has bounded trajectories on compact intervals. Hence the desired result.

c. passing to the limit

We can assert that

$$\tilde{E}^{Z^s}\left[\varepsilon\eta^2\frac{(2+\varepsilon\eta)}{(1+\varepsilon\eta)^2}(1+|x_s|^2)^{\frac{1}{2}}\right]\to 0, \text{ a.s. a.e.} \tag{2.2.5}$$

at least for a subsequence, since we have convergence in L^1. Therefore, also

$$\text{a.s. } \tilde{E}^{Z^s}\left[\varepsilon\eta^2\frac{(2+\varepsilon\eta)}{(1+\varepsilon\eta)^2}\left(1+|x_x|^2\right)^{\frac{1}{2}}\right]\to 0, \text{ a.e. t.} \tag{2.2.6}$$

Since also

$$\tilde{E}^{Z^s}\left[\varepsilon\eta^2\frac{(2+\varepsilon\eta)}{(1+\varepsilon\eta)^2}\left(1+|x_s|^2\right)^{\frac{1}{2}}\right]\leq 2\tilde{E}^{Z^s}\left[\eta(1+|x_t|^2)^{\frac{1}{2}}\right]$$

which belongs to $L^\infty(0,\, T)$, we deduce from (2.2.6) that

$$\int_o^T |\tilde{E}^{Z^s}\left[\varepsilon\eta^2\frac{(2+\varepsilon\eta)}{(1+\varepsilon\eta)^2}(1+|x_s|^2)^{\frac{1}{2}}\right]|^2 ds\to 0. \tag{2.2.7}$$

But, (2.2.7) implies

$$\int_o^T |\tilde{E}^{Z^t}\left[\Psi(\eta(1+\varepsilon\eta)^{-2}-\eta)h^*\right]|^2 ds\to 0. \tag{2.2.8}$$

From the theory of stochastic integral, we deduce from (2.2.8) that

$$\sup_{0\leq t\leq T}|\int_o^t \tilde{E}^{Z^t}\cdot\left[\Psi\eta((1+\varepsilon\eta)^{-2}-1)h^*\right]R^{-1}dz|\to 0 \tag{2.2.9}$$

in probability.

It is then possible to pass to the limit in the stochastic integral appearing at the right hand side of (2.2.1). This concludes the proof of Theorem 2.1.1.

Remark 2.2.1 From the equality (2.1.16), we deduce that $\int_o^t p(s)(h^*(s)\Psi(s))$ $R_s^{-1}dz$ belongs to $L^1(\Omega,\, Z^t,\, \tilde{P})$. Moreover, from the proof made above, it has mean 0.

It is very useful to give the following integrated form of Theorem 2.1.1.

Integrated form of Theorem 2.1.1. We have the following relation

$$\tilde{E}\theta_t p(t)(\Psi(t))=\pi_o(\Psi(0))$$

$$+\tilde{E}\int_o^t \theta_s p(s)\left(\frac{\partial\Psi}{\partial s} - A(s)\Psi(s) - i\Psi(s)h^*(s)R_s^{-1}\beta(s)\right) ds,\ \forall\, \beta(.). \tag{2.2.10}$$

This relation is a trivial consequence of the statement of Theorem 2.1.1, when h is bounded. Otherwise, it is proven directly from the limit procedure. Similarly, we can give the following

Integrated form of Theorem 2.1.2

$$E\theta_t\pi(t)(\Psi(t)) = \pi_o(\Psi(0))$$

$$+E\int_o^t \theta_s\pi(s)\left(\frac{\partial\Psi}{\partial s} - A(s)\Psi(s) - i\Psi(s)h^*(s)R_s^{-1}\beta(s)\right) ds,\ \forall\beta(.). \tag{2.2.11}$$

2.2.2 Proof of Theorem 2.1.2

We sketch it for the general case of unbounded coefficients. Firstly, from Zakai equation (2.1.16) applied with $\Psi = 1$, it follows that

$$\begin{aligned} p(t)(1) &= 1 + \int_o^t p(s)(h^*(s))R_s^{-1}dz \\ &= 1 + \int_o^t p(s)(1)\pi(s)(h^*(s))R_s^{-1}dz. \end{aligned}$$

Applying Ito's formula to $\log\left[\varepsilon + (p(t)(1))^2\right]^{\frac{1}{2}}$ and passing to the limit, we can prove that

$$p(t)(1) = \exp\left\{\int_o^t \pi(s)(h^*(s))R_s^{-1}dz - \frac{1}{2}\int_o^t \pi(s)(h^*(s))R_s^{-1}\pi(s)(h(s))ds\right\}.$$

Hence, we can consider $\frac{1}{p(t)(1)}$ and its Ito differential $d\frac{1}{p(t)(1)}$. Note that we make a constant use of the fact that $p(t)(1)$ and $\frac{1}{p(t)(1)}$ belong to $L^\infty(0, T)$, a.s. From these considerations, it follows that we may proceed as in the case of bounded coefficients.

2.3 UNIQUENESS

We shall consider in this section the problem of uniqueness of the Zakai and Kushner equation.

2.3.1 Statement of the result for Zakai equation

In fact, we shall prove the uniqueness of the solution of the integrated form of Zakai equation in a convenient functional space. This functional space is made precise as follows p(t) is a process with values in the space of finite positive measures on R^n.

If Ψ is Borel bounded then $(p(t)(\Psi)) \in L^1(\Omega,\, Z^t,\, \tilde{P}), \tilde{E}|p(t)(\Psi)| \leq ||\Psi||$.

If $\Psi(x,t)$ is Borel and a.e. t $|\Psi(x,t)| \leq K(1+|x|^2)$, then $p(t)(\Psi(t))$ is defined and a.e. t belongs to $L^1(\Omega,\, Z^t,\, \tilde{P})$.

If $\Psi_k(x)$ satisfies $||\Psi_k|| \leq C$, and $\Psi_k(x) \to \Psi(x)$ pointwise, then $\forall\, T$, $p(t)(\Psi_k) \to p(t)(\Psi)$ a.s.

If $\Psi_k(x,\, t)$ is Borel and a.e. t $|\Psi_k(x,t)| \leq K(1+|x|^2)$ $\forall x$ and $\Psi_k \to \Psi$ pointwise, then a.e. t $p(t)(\Psi_k(t)) \to p(t)(\Psi(t))$ a.s.

(2.3.1)

We can then state the

Theorem 2.3.1 We make the assumptions of Theorem 2.1.1 and (2.3.8) below. Then there exists a unique solution of the integrated form of Zakai equation (2.2.10), in the functional space defined by (2.3.1).

Before proving Theorem 2.3.1, we can remark that the process

$$p(t)(\Psi) = \tilde{E}\left[\Psi(x_t)\eta_t | Z^t\right]$$

belongs to the functional space defined by (2.3.1) and thus is the unique solution of Zakai equation, satisfying simultaneously (2.2.10) and (2.1.16).

From the integrated from of Zakai equation, it follows that there is a motivation to study the partial differential equation (P.D.E.)

$$\frac{\partial v}{\partial t} - A(t)v + ivh^* R^{-1}\beta = 0 \qquad v(x,\, T) = \phi(x) \tag{2.3.2}$$

where ϕ is a given Borel function (complex valued). In fact, (2.3.2) is a system, which reads

$$\begin{aligned} &\frac{\partial v_1}{\partial t} - A(t)v_1 - v_2 h^* R^{-1}\beta = 0, \;\; v_1(x,\, T) = \phi_1(x) \\ &\frac{\partial v_2}{\partial t} - A(t)v_2 - v_1 h^* R^{-1}\beta = 0, \;\; v_2(x,\, T) = \phi_2(x) \end{aligned} \tag{2.3.3}$$

in which $v = v_1 + iv_2$, $\phi = \phi_1 + i\phi_2$.

Let us solve (2.3.3) in the linear case

$$g(x,\, t) = F(t)x + f(t), \quad \sigma(x,\, t) = \sigma(t), \quad h(x,t) = H(t)x + h(t) \tag{2.3.4}$$

where we have taken

$$\phi(x) = \exp i\lambda^* x. \tag{2.3.5}$$

The solution of (2.3.2) is of the form

$$v(x,\, t) = \exp(i\rho^*(t)x + s(t)) \tag{2.3.6}$$

where $\rho(t),\ s(t)$ are identified by

$$\dot{\rho} + F^*\rho + H^*R^{-1}\beta = 0 \qquad \rho(T) = \lambda$$

$$\dot{s} + i\rho^* f + i\beta^* R^{-1} h - \frac{1}{2}\rho^*\sigma Q\sigma^*\rho = 0, \qquad s(T) = 0 \tag{2.3.7}$$

Turning to the general case, we are interested in getting smooth solutions $v_1,\ v_2$ of (2.3.3). We can prove the following

Lemma 2.3.1 Assume that

$$a_{ij,\lambda},\ a_{ij,\lambda\mu},\ g_{i,\lambda},\ g_{i,\lambda\mu},\ h_{i,\lambda},\ h_{i,\lambda\mu} \text{ are bounded.} \tag{2.3.8}$$

Then there exists a unique solution of (2.3.2) such that $v_1,\ v_2;\ v_{1,\lambda};\ v_{2,\lambda};\ v_{1,\lambda\mu},\ v_{2,\lambda\mu}$ are bounded.

Sketch of the proof. We define successively

$$z = \frac{1}{2}(v_1^2 + v_2^2), \qquad z = \sum_\lambda \frac{1}{2}(v_{1\lambda}^2 + v_{2\lambda}^2)$$

$$z = \frac{1}{2}\sum_{\lambda\mu}(v_{1\lambda\mu}^2 + v_{2\lambda\mu}^2)$$

and compute $-\frac{\partial z}{\partial t} + A(t)z$. In the first case we get $-\frac{\partial z}{\partial t} + A(t)z \leq 0$. In the two other cases, we get $-\frac{\partial z}{\partial t} + A(t)z \leq c_o z + c_1$.

From Gronwall's inequality, the desired result follows.

We now proceed with the proof of Theorem 2.3.1. We notice that the solution of (2.3.2) satisfies also

$$|\frac{\partial v}{\partial t}| \leq C(1 + |x|)$$

which follows from the equation itself.

We pick a sequence v_k of functions in $C_b^{2,1}$ such that

$$|v_k(x,\,t)|, \qquad |Dv_k(x,\,t)| \qquad |D^2v_k(x,t)| \leq C, \quad |\frac{\partial v_k}{\partial t}(x,\,t)| \leq C(1+|x|)$$

and v_k and its derivatives converge pointwise.

From the integrated form of Zakai equation, we can write

$$\tilde{E}\ \theta(T)p(T)(v_k(T)) = \pi_o(v_k(0)) + \tilde{E}\int_o^T \theta(t)p(t)\left(\frac{\partial v_k}{\partial t} - A(t)v_k + iv_k h^* R^{-1}\beta\right) dt.$$

Using the properties of the functional space (2.3.1), one can pass to the limit in the above expression and obtain

$$\tilde{E}\ \theta(T)p(T)(\phi) = \pi_o(v(0)).$$

Therefore, if $p_1(t),\ p_2(t)$ are two solutions necessarily

$$\tilde{E}\ \theta(T)p_1(T)(\phi) = \tilde{E}\ \theta(T)p_2(T)(\phi).$$

and using Lemma 2.1.3 it follows that

$$p_1(T)(\phi) = p_2(T)(\phi).$$

This relation extends to functions ϕ which are continuous and bounded. This suffices to infer that $p_1(T) = p_2(T)$.

2.3.2 Uniqueness of the solution of Kushner equation

We need to make precise the functional space in which we look for the solution of (2.1.20). It is the following

$\pi(t)$ is a process with values in the space of probability measures on R^n.
If Ψ is Borel bounded, then $\pi(t)(\Psi)$ is Z^t measurable and $|\pi(t)(\Psi)| \leq ||\Psi||$ a.s. then
$|\pi(t)(\Psi)|^2 \leq \pi(t)(\Psi^2)$ a.s.
$\pi(t)$ extends to functions $\Psi(x,t)$ which are Borel and satisfy
$|\Psi(x,t)| \leq K(1+|x|^2))$ a.e. t, $\forall x$.

Then a.e. t $\pi(t)(\Psi(t)) \in L^1(\Omega, Z^t, \tilde{P})$. If $\Psi_k(x)$ satisfy $||\Psi_k|| \leq C$, $\Psi_k(x) \to \Psi(x)$ pointwise then $\forall t, \pi(t)(\Psi_k) \to \pi(t)(\Psi)$ a.s.

If $\Psi_k(x, t)$ is Borel and satisfies a.e. t $|\Psi_k(x, t)| \leq K(1 + |x|^2) \forall x$,

and $\Psi_k \to \Psi$ pointwise, then one has

a.e. t, $\pi(t)(\Psi_k(t)) \to \pi(t)(\Psi(t))$ a.s.

(2.3.9)

We can then state the

Theorem 2.3.2 We make the assumptions of Theorem 2.3.1. Then there exists a unique solution of Kushner equation (2.1.20), in the functional space defined by (2.3.9).

Sketch of the proof. The steps are the following. For any solution π of (2.1.20), define

$$\zeta_t = \exp\left\{\int_o^t \pi(s)(h^*(s))R_s^{-1}dz - \frac{1}{2}\int_o^t \pi(s)(h^*(s))R_s^{-1}\pi(s)(h(s))ds\right\}$$

and

$$p(t)(\Psi(t)) = \pi(t)(\Psi(t))\zeta_t.$$

The main point is to prove that $p(t)(\Psi(t))$ is solution of the integrated form of Zakai equation in the functional space (2.3.1). In that case $\zeta_t = p(t)(1)$, and thus $\pi(t)(\Psi(t))$ is uniquely defined.

Let us mention some details.

a. We first prove estimates

Let $\chi(x) = 1 + |x|^2$, then one has

$$E|x_t|^2 \rho_t \zeta_t \leq C \tag{2.3.10}$$

$$E\pi(t)(\chi)\rho_t\zeta_t \leq C. \tag{2.3.11}$$

These estimates are proven by considering

$$\chi_\lambda = \frac{1 + |x|^2}{1 + \lambda|x|^2}$$

and making use of Ito's formula to compute the quantity $\left(\frac{\pi(t)(\chi_\lambda)\rho_t\zeta_t}{1+\varepsilon\pi(t)(\chi_\lambda)\rho_t\zeta_t}\right)$. We use the properties of $\pi(t)$ arising from the functional space (2.3.9) to pass to the limit. The estimate (2.3.11) then follows from Fatou's lemma.

b. We prove that

$$\tilde{E}\zeta_t = 1. \tag{2.3.12}$$

This amounts to proving that

$$E\rho_t\zeta_t = 1.$$

We apply Ito's formula to $\frac{\rho_t\zeta_t}{1+\varepsilon\rho_t\zeta_t}$.

We use the estimate (2.3.11) to pass to the limit.

c. Verification of (2.3.1)

The main point is that if $|\Psi(x,t)| \leq K(1+|x|^2) = K\chi$, then

$$\tilde{E}|p(t)(\Psi(t))| \leq K\, E\,\rho_t\zeta_t(\chi) < \infty.$$

d. Verification of the integrated form of Zakai equation

This amounts to showing the relation

$$E\,\pi(t)(\Psi(t))\theta_t\rho_t\zeta_t = \pi_0(\Psi(0)) + E\int_0^t \theta_s\rho_s\zeta_s\pi(s)\left(\frac{\partial\Psi}{\partial s} - A(s)\Psi + ih^*\beta\Psi\right)ds. \tag{2.3.13}$$

The method consists in applying Ito's formula to compute $\frac{\pi(t)(\Psi(t))\theta_t\rho_t\zeta_t}{1+\varepsilon\rho_t\zeta_t}$.

2.3.3 Linear case

As an example, let us solve Zakai and Kushner equations in the linear case, where

$$g(x,t) = F(t)x + f(t)$$

$$\sigma(x,t) = I$$

$$h(x,t) = H(t)x + h(t).$$

The assumptions ensuring the existence and uniqueness of the solutions of Zakai and Kushner equations are satisfied. If we take π_o such that

$$\pi_o(\Psi) = \int \Psi\left(x_o + P_o^{\frac{1}{2}}\xi\right)\frac{\exp -\frac{1}{2}|\xi|^2}{(2\pi)^{n/2}}d\xi$$

then the unique solution of Zakai equation is given by

$$p(t)(\Psi) = s_t \int \Psi(\hat{x}_t + P_t^{1/2}\xi)\frac{\exp -\frac{1}{2}|\xi|^2}{(2\pi)^{n/2}}d\xi$$

where p_t is the solution of

$$P + PH^*R^{-1}HP - Q - FP - PF* = 0 \qquad P(0) = Po$$

and $\hat{x}_t$ is the Kalman filter,

$$d\hat{x}_t = (F\hat{x}_t + f)dt + PH^*R^{-1}(dz - (H\hat{x} + h)dt) \qquad \hat{x}(0) = x_o.$$

Moreover, s_t is defined by

$$s_t = \exp\left\{\int_o^t (\hat{x}^*H^* + h^*)R^{-1}dz - \frac{1}{2}\int_o^t (\hat{x}^*H^* + h^*)R^{-1}(H\hat{x} + h)ds\right\}$$

and necessarily $s_t = p(t)(1)$.

Remark 2.3.1 Kalman filter theory tells only that

$$\hat{x}_t = E[x_t|Z^t]$$
$$P_t = E[(x_t - \hat{x}_t)(x_t - \hat{x}_t)^*|Z^t]$$

whereas we have obtained the conditional probability itself by solving Zakai equation.

2.4 CORRELATION BETWEEN THE SIGNAL NOISE AND THE OBSERVATION NOISE

In the previous model, there was no correlation between the signal noise and the observation model. We consider a more general case in this section.

2.4.1 A simplified model with correlation

The model is described as follows

$$dx_t = g(x_t, t)dt + \sigma(x_t, t)dw_t$$
$$x(0) = \xi \tag{2.4.1}$$
$$dz_t = h(x_t, t)dt + \alpha(t)dw_t + db_t$$
$$z(0) = 0. \tag{2.4.2}$$

The assumptions on g, σ, h are the same as those stated in §1.1. Moreover,

$$\alpha(t) \text{ is a deterministic bounded matrix.} \tag{2.4.3}$$

We set

$$D = R + \alpha Q \alpha^*$$

and define ρ_t by

$$d\rho = -\rho h^*(x_t) D^{-1}(\alpha dw + db), \quad \rho(0) = 1.$$

Let next $\tilde{P}$ be a new probability on Ω, A, whose Radon Nikodym derivative with respect to P is given by

$$\frac{d(\tilde{P})}{dP} \mid F^t \ = \ \rho_t$$

and let $\eta_t = 1/\rho_t$.

We consider also the process

$$\tilde{w}(t) = w(t) + \int_0^t Q\alpha^* D^{-1} h(x_s) ds.$$

We then have the following result

Lemma 2.4.1 For $(\Omega, A, \tilde{P}, F^t)$ the process $\tilde{w}$ and z are F^t Wiener processes with covariance Q and D and correlation.

$$\tilde{E}[(z(t) - z(s))(\tilde{w}(t) - \tilde{w}(s))^*] = \int_s^t \alpha Q(\tau) d\tau. \tag{2.4.4}$$

We can also notice that for $(\Omega, A, \tilde{P}, F^t)$ the process $x(.)$ is the solution of

$$\begin{aligned} dx_t &= [g(x_t) + \sigma(x_t) Q_t \sigma_t^* D_t^{-1} h(x_t)]dt + \sigma(x_t) d\tilde{w}_t \\ x(0) &= \xi. \end{aligned} \tag{2.4.5}$$

We can then state the

Theorem 2.4.1 We make the assumptions of Theorem 2.1.1 and (2.4.3), then the unnormalized conditional probability is solution of the Zakai equation.

$$p(t)(\Psi(t)) = \pi_0(\Psi(0)) + \int_0^t p(s)\left(\frac{\partial \Psi}{\partial s} - A(s)\Psi(s)\right) ds$$

$$+\int_o^t p(s)\left[h^*(s)\Psi(s)+D\Psi^*(s)\sigma(s)Q(s)\alpha^*(s)\right]D_s^{-1}dz \text{ a.s.}, \quad \forall t. \tag{2.4.6}$$

To establish the uniqueness we need to study a system of P.D.E. analogue to (2.3.3) It is the following

$$\begin{aligned} &\frac{\partial v_1}{\partial t}-A(t)v_1-\beta^*D^{-1}(hv_2+\alpha Q\sigma^*Dv_2)=0 \\ &\frac{\partial v_2}{\partial t}-A(t)v_2+\beta^*D^{-1}(hv_1+\alpha Q\sigma^*Dv_1)=0 \end{aligned} \tag{2.4.7}$$

$$v_1(s,T)=\phi_1(x),\ v_2(x,\,T)=\phi_2(x).$$

We can obtain the same regularity results as in the case $\alpha=0$ (Lemma 2.3.1), provided we make the additional assumptions

σ is bounded; a_{ij} is strongly elliptic.

Similarly, we can establish the

Theorem 2.4.2 We make the assumptions of Theorem 2.1.1 and (2.4.3), then the normalized conditional probability is given by

$$\begin{aligned} \pi(t)(\Psi(t)) &= \pi_0(\Psi(0))+\int_o^t \pi(s)(\frac{\partial\Psi}{\partial s}-A(s)\Psi(s))ds \\ &+\int_o^t [\pi(s)(h^*(s)\Psi(s)+D\Psi^*(s)\sigma(s)Q(s)\alpha^*(s) \\ &-\pi(s)(\Psi(s))\pi(s)(h^*(s))]D_s^{-1}(dz-\pi(s)(h(s))ds). \end{aligned} \tag{2.4.9}$$

As an example, we can consider the linear case and obtain the following formulas

$$p(t)(\Psi)=s_t\int\Psi\left(\hat{x}_t+P_t^{1/2}\xi\right)\frac{\exp-\frac{1}{2}|\xi|^2}{(2\pi)^{n/2}}d\xi \tag{2.4.10}$$

where

$$\dot{P}+(PH^*+Q\alpha^*)D^{-1}(HP+\alpha Q)-Q-FP-PF^*=0 \qquad P(0)=P_o \tag{2.4.11}$$

and the Kalman filter $\hat{x}$ is the solution of

$$\begin{aligned} &d\hat{x}_t=(F\hat{x}_t+f)dt+(PH^*+Q\alpha^*)D^{-1}(dz-(H\hat{x}+h)dt) \\ &\hat{x}(0)=x_o \end{aligned} \tag{2.4.12}$$

and $s_t=p(t)(1)$ is given by

$$s_t=\exp\left\{\int_o^t(\hat{x}^*H^*+h^*)D^{-1}dz-\frac{1}{2}\int_o^t(\hat{x}^*H^*+h^*)D^{-1}(H\hat{x}+h)ds\right\}. \tag{2.4.13}$$

2.4.2 Further correlation

We consider a more stringent correlation between the signal noise and the observation noise. The model is the following

$$\begin{aligned} dx_t &= g(x_t,\ z_t)dt + \sigma(x_t, z_t)dw \\ x(0) &= \xi \\ dz_t &= h(x_t, z_t)dt + \alpha(z_t)dw + db \\ z(0) &= 0 \end{aligned} \tag{2.4.14}$$

The only restriction that remains is that α does not depend on x. The case when α is a function of x is still an open problem.

Let us define

$$D(z) = R + \alpha(z)Q\alpha^*(z)$$

$$B(z) = Q - Q\alpha^*(z)D(z)^{-1}\alpha(z)Q$$

and note that $B(z)$ is positive definite.

We define ρ_t to be the solution of

$$d\rho_t = -\rho_t h^*(x_t, z_t)D(z_t)^{-1}(\alpha dw + db),\ \rho(0) = 1$$

We cannot keep any more z_t as a Wiener process. We define instead $\tilde{z}_t, \tilde{w}_t$ by the relations

$$d\tilde{z}_t = D(z_t)^{-\frac{1}{2}}dz_t \qquad \tilde{z}(0) = 0 \tag{2.4.15}$$

$$d\tilde{w}_t = B(z_t)^{-\frac{1}{2}}z_t - [dw - Q\alpha^*(z_t)D(z_t)^{-1}dz_t + Q\alpha^*(z_t)D(z_t)^{-1}h(x_t, z_t)dt]$$

$$\tilde{w}(0) = 0 \tag{2.4.16}$$

and we perform the change of probability

$$\frac{d\tilde{P}}{dP}\Big|_{F^t} = \rho_t.$$

We can then state the

Lemma 2.4.2 For the system $(\Omega,\ A,\ \tilde{P},\ F^t)$ the processes $\tilde{w}_t$ and $\tilde{z}_t$ are independent standard Wiener processes.

Sketch of the proof. Consider the stochastic integral

$$I_t(\phi, \Psi) = \int_t^T \phi^* d\tilde{z} + \int_t^T \Psi^* d\tilde{w}$$

where ϕ, ψ are L^2 deterministic functions. We can check that

$$\tilde{E}[\exp iI_t(\phi,\psi)\ F^t] = \exp -\frac{1}{2}\int_t^T (|\phi|^2 + |\psi|^2)ds$$

and this implies the desired result.

Moreover, for the system $(\Omega,\ A,\ \tilde{P}, F^t)$ the processes x_t and z^t appear as the solutions of

$$dx_t = [g(x_t,z_t) - \sigma(x_t,z_t)Q\alpha^*(z_t)D(z_t)^{-1}h(x_t,z_t)]dt$$

$$+\sigma(x_t,z_t)Q\alpha^*(z_t)D(z_t)^{-\frac{1}{2}}d\tilde{z}_t + \sigma(x_t,z_t)B(z_t)^{\frac{1}{2}}d\tilde{w}_t \tag{2.4.17}$$

$$x(0) = x_0,\ z(0) = 0. \tag{2.4.18}$$

If $D(z)$ is Lipschitz, we can then assert that $Z^t = \tilde{Z}^t$. Let $\eta_t = 1/\rho_t$. If $\chi(x,z)$ is a Borel bounded test function, we shall define

$$p(t)(\chi) = \tilde{E}[\xi(x_t,z_t)\eta_t|Z^t]. \tag{2.4.19}$$

This definition extends the usual one, in which we consider test functions, which depend on x only.

We obtain the following result, which is stated without making precise all the assumptions on g, σ, h: the unnormalized conditional probability $p(t)$ is the solution of the Zakai equation

$$p(t)(\Psi(t)) = \pi_0(\Psi(0)) + \int_0^t p(s)\left(\frac{\partial\Psi}{\partial s} - A(s)\Psi\right)ds$$

$$+\int_0^t [p(s)(D\Psi^*(s)\sigma(s))Q_s\alpha^*(z_s) + p(s)(\Psi(s)h^*(s))]\ D(z_s)^{-1}dz_s. \tag{2.4.20}$$

In (2.4.20) the test function $(\Psi(x,t))$ depends only on x, as in the previous statements of Zakai equation. However, it is important in this context to consider test functions which depend on both x and z. This is the only way to derive a uniqueness result.

If we consider a test function $\Psi(x,z,t)$ which is $C^{2,2,1}(R^n \times R^m \times [0,\infty])$ then one has the relation

$$\begin{aligned} p(t)(\Psi(t)) &= \pi_0(\Psi(.,0,0,)) \\ &+ \int_0^t p(s)(\frac{\partial \Psi}{\partial s} - A\Psi + h^* D_Z \Psi + \frac{1}{2} tr\ D(z) D_z^2 \Psi) ds \\ &+ \int_0^t [p(s)(D_x \Psi^*(s)\sigma(s))Q_s \alpha^*(z_s) \\ &+ p(s)(\Psi(s)h^*(s)) + p(s)(D_z\Psi^*(s))D(z_s)] D(z_s)^{-1} dz_s. \end{aligned} \tag{2.4.21}$$

To derive a uniqueness result one can consider the complex valued P.D.E.

$$\frac{\partial v}{\partial t} - A(t)v + h^* D_z v + \frac{1}{2} tr D(z) D_z^2 v + i\beta^* D^{-1}(hv + DD_z v + \alpha Q \sigma^* D_x v) = 0$$

$$v(x,z,T) = \phi(x). \tag{2.4.22}$$

When the coefficients do not depend on z, this equation has a unique solution which does not depend on z, and we recover (2.3.2).

2.5 REPRESENTATION FORMULAS FOR THE UNNORMALIZED DENSITY

We present in this section some important representation formulas for the unnormalized density, which is the density of the unnormalized conditional probability with respect to Lebesgue's measure, when of course it exists.

We follow the work of J. Picard [8] and E. Pardoux [7].

2.5.1 Existence of an unnormalized density

To simplify the problem, we restrict ourselves to the case

$$dx_t = g(x_t,t)dt + \sigma(x_t,t)dw_t, \qquad x(0) = \xi$$

$$dz_t = h(x_t,t)dt + db_t \qquad z(0) = 0 \tag{2.5.1}$$

where

$$g,\ \sigma,\ h \text{ are bounded} \tag{2.5.2}$$

$$g_x,\ \sigma_x \text{ are bounded} \tag{2.5.3}$$

π_0 has a density with respect to Lebesgue's measure and $\int |x|^2 p_0(x)dx < \infty$

(2.5.4)

w, b are independent standard Wiener processes; ξ is a random variable independent of w, b

(2.5.5)

$a(x,t) \leq \delta I,\ \delta > 0.$

(2.5.6)

If there is a density, $p(x,t)$ we can write

$$p(t)(\Psi) = \int p(x,t)\Psi(x)dx$$

and from Zakai equation, we deduce

$$d\int p(x,t)\Psi(x)dx = \int p(x,t)\left(\frac{\partial \Psi}{\partial t} - A(t)\Psi\right)dt + \left[\int p(x,t)\Psi(x,t)h^*(x,t)dx\right]dz.$$

Performing a formal integration by parts in the 1st integral to the right of the preceding equality, we see that p must satisfy the equation

$$dp + A^*(t)pdt = ph^*dz \qquad p(x,0) = p_o(x). \tag{2.5.7}$$

The equation (2.5.7) is written very formally. We shall perform an additional manipulation to recover an ordinary P.D.E.

2.5.2 The robust form

To proceed, we need some additional regularity assumptions on the function h, namely

$$\frac{\partial h}{\partial t}, Dh,\ D^2h \text{ bounded.} \tag{2.5.8}$$

We perform on (2.5.7) the change of unknown function

$$q(x,t) = p(x,t)\exp -h^*(x,t)z(t). \tag{2.5.9}$$

The interest of this transformation is that we can get rid of the Ito differential and recover an ordinary P.D.E. (with random coefficients). Indeed q is the solution of

$$\frac{\partial q}{\partial t} + A^*q - 2aDq.D(hz) + qe_t(h,z) = 0$$

$$q(x,0) = p_0(x) \tag{2.5.10}$$

where we have made use of the following notation

$$e_t(h,z) = \frac{\partial h}{\partial t}\cdot z_t + A(t)(h_t.z_t) + 2f_t.D(h_t.z_t) - a(t)D(h_t.z_t)D(h_t.z_t) + \frac{1}{2}|h|^2, \tag{2.5.11}$$

in which the vector f_t has the components

$$f_i(t) = g_i(t) - a_{ij,j}(t).$$

In (2.5.11) we have used the notation of scalar product instead of the matrix notation for line vectors. Thus comparing to previous notation $h.z_t = h^* z_t$ and similar definitions.

Now (2.5.10) is an ordinary parabolic P.D.E., whose coefficients depend parametrically on $z(.)$ (and therefore are random). We can then study (2.5.10) directly, whenever for instance

$$p_o \in L^2(R^n). \tag{2.5.12}$$

From the standard variational theory for parabolic P.D.E., we deduce that there exists a unique solution of (2.5.10) q such that

$$\text{a.s.} \quad q \in L^2(0,T;\, H^1(R^n)), \frac{\partial q}{\partial t} \in L^2(0,T; H^{-1}(R^n)). \tag{2.5.13}$$

The assumptions (2.5.2), (2.5.3), (2.5.6) suffice to imply the result (2.5.13). We may then define $p(x,t)$ by (2.5.9). The process p satisfies also (2.5.13).

This is however not sufficient to prove that p satisfies (2.5.7). In particular, we do not know if p has moments. A direct study of (2.5.7) is necessary. But for the representation formula that we have in mind, it is sufficient to define p through (2.5.9) and (2.5.10).

First representation formula. We first define a notation. If $\phi(x,t)$ is any function, we shall write

$$\bar{\phi}(x,t) = \phi(x, T-t).$$

Similarly, we write

$$\bar{z}_t = z(T-t), \bar{A}(t) = A(T-t),$$

$$\bar{e}_t(\bar{h},\bar{z}) = -\frac{\partial \bar{h}}{\partial t}\bar{z}_t + \bar{A}(t)(\bar{h}_t.\bar{z}_t) + 2\bar{f}_t.D(\bar{h}_t.\bar{z}_t) - \bar{a}(t)D(\bar{h}_t.\bar{z}_t) + \frac{1}{2}|\bar{h}_t|^2.$$

The $\bar{q}(x,t)$ is the solution of

$$-\frac{\partial \bar{q}}{\partial t} + \bar{A}^*\bar{q} - 2\bar{a}D\bar{q}.D(\bar{h}.\bar{z}) + \bar{q}\bar{e}_t(\bar{h},\bar{z}) = 0$$

$$\bar{q}(x,T) = p_o(x) \tag{2.5.14}$$

Let us introduce

$$\tilde{q}(x,t) = \bar{q}(x,t)\exp\bar{h}(x,t)z_T \tag{2.5.15}$$

and the process

$$\tilde{z}_t = z_T - z_t. \tag{2.5.16}$$

We note that $\tilde{z}_t$ is a standard Wiener process with respect to $\tilde{Z}^t = \sigma(\tilde{z}_s, s \le t)$, and $\tilde{z}_T = z_T$, $\tilde{z}_0 = 0$. Next, one can check that $\tilde{q}$ is the solution of

$$-\frac{\partial \tilde{q}}{\partial t} + \bar{A}\tilde{q} + (2\bar{a}D(\bar{h}.\tilde{z}) + 2\bar{f}).D\tilde{q} + \tilde{q}[\text{div}\bar{f} + \tilde{e}_t(\bar{h}.\tilde{z})] = 0$$

$$\tilde{q}(x,T) = p_0(x)\exp\bar{h}(x,T).\tilde{z}_T \tag{2.5.17}$$

where

$$\tilde{e}_t(\bar{h}.\tilde{z}) = \frac{\partial \bar{h}}{\partial t}.\tilde{z}_t - \bar{A}(t)(\bar{h}_t.\tilde{z}_t) - 2\bar{f}_t.D(\bar{h}_t.\tilde{z}_t) - \bar{a}_tD(\bar{h}_t.\tilde{z})D(\bar{h}_t.\tilde{z}) + \frac{1}{2}|\bar{h}_t|^2.$$

It is very important to notice that the coefficients of the equation (2.5.17) depend only on $\tilde{z}$.

The main point consists in giving a probabilistic interpretation of the solution of (2.5.17), as an ordinary backward parabolic equation, using an additional probability space independent of $\tilde{Z}$.

Consider another probability space Ω_0, A_0, P_0 with a filtration of Γ^t and a standard Wiener process w_{ot} (which is a Γ^t Wiener) with values in R^n.

We may consider as probabilistic set up $\Omega \times \Omega_0, A \otimes A_0, \tilde{P} \otimes P_0 = Q$ equipped with the filtration $\tilde{Z}^t \otimes \Gamma^t$ (we notice that the pair $\tilde{z}_t, w_{ot}$ is a pair of independent Wiener processes). Consider the Ito equation

$$dX = (\bar{g} - 2\bar{f})(X)dt + \bar{\sigma}(X)dw_o$$

$$X(0) = x \tag{2.5.18}$$

where x is given.

Define the process ζ_t by

$$d\zeta_t = \zeta_t\{(-\text{div}\bar{f}(X_t) - \tilde{e}_t(\bar{h},\tilde{z})(X_t))dt - D(\bar{h}.\tilde{z})\bar{\sigma}(X_t)dw_0\} \qquad \zeta(0) = 1 \tag{2.5.19}$$

In the equation (2.5.19) $\tilde{z}$ is considered as a parameter, the randomness is due to w_0 (and thus also through X). By the very definition, $\tilde{q}(X_t,t)\zeta_t$ is a P_0,Γ^t martingale and

$$E_0\tilde{q}(X_T,T)\zeta_T = \tilde{q}(x,0)$$

i.e.

$$\tilde{q}(x,0) = E_0[p_0(X_T)\zeta_T \exp \bar{h}(X_T,T)\tilde{z}_T].$$

We now replace ζ_T by its value obtained from (2.5.19), and take account of the independence of X_t and $\tilde{z}_t$ to write

$$\begin{aligned}\tilde{q}(x,0) = E^Q[p_0(X_T)(\exp - \int_0^T \text{div}\bar{f}(X_t)dt)\exp\{\bar{h}(X_T,T).\tilde{z}_T \\ + \int_0^T [-\frac{\partial\bar{h}}{\partial t}.\tilde{z}_t + \bar{A}(t)(\bar{h}_t.\tilde{z}_t) + 2\bar{f}_t.D(\bar{h}_t.\tilde{z}_t) - \frac{1}{2}|\bar{h}_t|^2]dt \\ - \int_0^T D(\bar{h}_t.\tilde{z}_t)\bar{\sigma}dw_0\}\tilde{Z}^T].\end{aligned} \tag{2.5.20}$$

Noting finally that

$$\tilde{q}(x,0) = p(x,T)$$

we obtain the formula, after some calculations

$$\begin{aligned}p(x,T) = E^Q[p_0(X_T)(\exp - \int_0^T \text{div}\bar{f}(X_t)dt) \\ \exp\{\int_0^T \bar{h}(X_T).d\tilde{z} - \frac{1}{2}\int_0^T |\bar{h}(X_t,t)|^2dt\}|\tilde{Z}^T].\end{aligned} \tag{2.5.21}$$

Remark 2.5.1 Applying the formula (2.5.21) with $h = 0$, we get the interpretation of the density of x_t which is the solution of the classical Fokker Planck equations.

We can avoid the introduction of the additional probabilistic set up Ω_0, A_0, P_0 and keep the intitial one. The role of the process X_t will be played now by

$\bar{x}_t = x(T-t)$. Let us be more specific. Consider the function $m(x,t)$ which is the density of x_T. It is the solution of

$$\frac{\partial m}{\partial t} + A^* m = 0 \qquad m(x,0) = p_0(x).$$

Let

$$\mathcal{X}^t = \sigma(\bar{x}_s, s \leq t).$$

It has been shown by U. Haussmann and E. Pardoux [2] that $\bar{x}_t$ is a diffusion. It appears as the solution of

$$d\bar{x}_t = [\bar{g} - 2\bar{f} + 2\bar{a}\ D \operatorname{Log} \bar{m}]\ (\bar{x}_t, t)dt + \bar{\sigma}(\bar{x}_t, t)d\ \hat{w}$$

$$\bar{x}(0) = x(T)$$

where $\hat{w}$ is a convenient $\mathcal{X}^t$ standard Wiener process. Consider on $\Omega, \bar{\mathcal{X}}^T, \otimes \tilde{Z}^T$ the probability

$$\frac{d\tilde{P}^T}{d\tilde{P}} = \exp\left\{\int_o^T [-D\ Log\ \bar{m} \cdot \bar{\sigma} d\hat{w} - \bar{a} D \operatorname{Log} \bar{m}.D \operatorname{Log} \bar{m}]\, dt\right\}$$

and define

$$w_0(t) = \hat{w}_t + \int_o^t \overline{\sigma^* D \operatorname{Log} m}(\bar{x}_s, s)ds.$$

It is then a standard $\Omega, \bar{\mathcal{X}}^T \otimes \tilde{Z}^T, \tilde{P}^T$ Wiener process adapted to the filtration $\bar{\mathcal{X}}^T$. Moreover, one has

$$d\bar{x} = (\bar{g} - 2\bar{f})(\bar{x}_t, t)dt + \bar{\sigma}(\bar{x}_t, t)dw_0, \qquad \bar{x}(0) = \xi \tag{2.5.22}$$

which is the equation used for X (although for a different Wiener process).

Collecting results, we obtain the formula

$$\begin{aligned} p(x,T) = \tilde{E}^T[p_0(\bar{x}_T)(\exp - \int_o^T \operatorname{div} \bar{f}(\bar{x}_t)dt) \\ \exp\{\int_o^T \bar{h}(\bar{x}_T)d\tilde{z} - \frac{1}{2}\int_o^T |\bar{h}(\bar{x}_t, t)|^2 dt\}|\tilde{Z}^T, \bar{x}_o = x] \end{aligned} \tag{2.5.23}$$

This formula has been established by J. Picard [8].

2.5.4 A second representation formula

Our objective in this paragraph is to give a representation formula for $p(T)(\Psi)$. We introduce the equation "dual" to (2.5.10)

$$-\frac{\partial v}{\partial t} + A(t)v + 2aDv.D(h,z) + vk_t(h,z) = 0, \qquad v(x,T) = \bar{v}(x) \tag{2.5.24}$$

where we have set

$$k_t(h,z) = \frac{\partial h}{\partial t} \cdot z_t - A(t)(h_t.z_t) - a(t)D(h_t.z_t).D(h_t.z_t) + \frac{1}{2}|h_t|^2. \tag{2.5.25}$$

The duality relation is made explicit by

$$\int p_0\ (x)\ v\ (x,o)\ dx = \int q(x,T)\ \bar{v}\ (x)\ dx. \tag{2.5.26}$$

If we pick

$$\bar{v}\ (x) = \Psi(x) \exp h\ (x,T).z_T$$

we obtain the formula

$$p(T)(\Psi) = \int p_0(x)v(x,0)dx. \tag{2.5.27}$$

Consider next the function

$$\tilde{v}(x,t) = v(x,t) \exp -h(x,t).z_T \tag{2.5.28}$$

and the process

$$\hat{z}_t = z_t - z_T. \tag{2.5.29}$$

This process differs from $\tilde{z}_t$. With these definitions, $\tilde{v}$ appears as the solution of

$$-\frac{\partial \tilde{v}}{\partial t} + A(t)\tilde{v} + 2aD\tilde{v}.D(h_t.\hat{z}_t) + \tilde{v}k_t(h,\hat{z}) = 0, \qquad \tilde{v}(x,T) = \Psi(x). \tag{2.5.30}$$

Consider $\hat{Z}^t = \sigma(\hat{z}_s, t \leq s \leq T)$, which is a decreasing family. We can then assert that the process $\hat{z}_t$ is a **backward Wiener process**. This means that it satisfies the properties

$$\tilde{E}(\hat{z}_{t_1} - \hat{z}_{t_2}|\hat{Z}^{t_2}) = 0 \quad \forall t_1 \leq t_2$$

$$\tilde{E}\left[(\hat{z}_{t_1} - \hat{z}_{t_2})(\hat{z}_{t_1} - \hat{z}_{t_2})^*|\hat{Z}^{t_2}\right] = (t_2 - t_1)|. \tag{2.5.31}$$

It is then possible to define a **backward stochastic integral** as follows. Let ϕ_t be adapted to $\hat{Z}^t$, we define

$$\int_o^T \phi_t.d\hat{z}_t = P.\lim \sum_i \phi_{t_i+1}(\hat{z}_{t_i+1} - \hat{z}_{t_i}).$$

It can be checked that $\int_t^T \phi_s.d\hat{z}_s$ is a $\hat{Z}^t$ backward martingale. One can also establish a backward Ito formula. Consider the process

$$\xi_t = \xi_T + \int_t^T a(s)ds + \int_t^T b(s).d\hat{z}_s \tag{2.5.32}$$

where $a(t), b(t)$ are $\hat{Z}^t$ adapted and

$$\int_o^T a(s)ds < \infty, \int_o^T b^2(s)ds < \infty \quad \text{a.s.}$$

Consider also $\Phi(x_t, T)$ which is $C^{2,1}$ then we can write the formula

$$\Phi(\xi_t, t) = \Phi(\xi_T, T) + \int_t^T \left(-\frac{\partial \Phi}{\partial s} + a(s)D\Phi + \frac{1}{2}trD^2\Phi bb^*\right) ds + \int_t^T D\Phi bd\hat{z}_s. \tag{2.5.33}$$

Let us then define

$$u(x,t) = \hat{v}(x,t)\exp -h(x,t).\hat{z}_t$$

which is adapted to $\tilde{Z}^t$.

Using Ito's backward stochastic calculus we get

$$-du + A(t)u\ dt = uh^*.d\hat{z}_t, \qquad u(x,T) = \Psi(x). \tag{2.5.34}$$

Note also that

$$u(x,0) = v(x,0).$$

Therefore, we obtain the representation formula (E. Pardoux [7])

$$p(T)(\Psi) = \int p_o(x)u(x,0)dx. \tag{2.5.35}$$

2.6 STOCHASTIC PARTIAL DIFFERENTIAL EQUATIONS

We have encountered two stochastic partial differential equations, namely (2.5.7) and (2.5.34) (forward and backward). They have been derived through the robust form.

This approach has two drawbacks. It does not carry over to the case when there is correlation between the signal and the observation noises. Besides, it requires unnecessary regularity on h and does not give information on expected values. We shall use in this section a direct approach, based on the work of E. Pardoux [7].

2.6.1 Setting of the model

Consider Zakai equation when there is correlation between the signal and observation noises (see (2.4.6)), and define

$$M_t(x) = M(x,t) = D_t^{-1}\alpha_t Q_t \sigma^*(x,t).$$

Using formal transposition after setting

$$p(t)(\Psi(t)) = \int p(x,t)\Psi(x,t)dx,$$

in a similar way as in § 2.5.1, we deduce

$$dp + A^* p\, dt = [-\mathrm{div}\,(pM^*) + ph^* D^{-1}]dz, \qquad p(x,0) = p_0(x). \tag{2.6.1}$$

The right hand side can be simplified when one has

$$M, \frac{\partial M_{ij}}{\partial x_j} \text{ bounded.} \tag{2.6.2}$$

In that case (2.6.1) yields

$$dp + A^* p dt = B^*(t) p dz, \qquad p(x,0) = p_0(x) \tag{2.6.3}$$

where

$$B^*(t)p(x) = p(x)\tilde{h}(x,t) - M(x,t)Dp \tag{2.6.4}$$

with

$$\tilde{h}_i(x,t) = (D^{-1}h)_i(x,t) - \sum_j \frac{\partial Mij}{\partial x_j}(x,t). \tag{2.6.5}$$

The notation $B^*(t)$ is to be considered in relation with another one denoted

$$B(t)v(x) = v(x)D_t^{-1}h(x,t) + M(x,t)Dv. \tag{2.6.6}$$

We shall consider $B(t)$ and $B^*(t)$ as operators in $L(V; H^m)$, where $V = H^1(R^n)$. The notation $*$ bears some ambiguity, since B^* is not the dual of B. However, this is reminiscent of some duality in other functional spaces, but we shall not make this explicit. We then pose the following problems: find solutions of the equations.

$$dp + A^* p dt = B^*(t) p dz$$

$$p(0) = p_0, p \in L^2_Z(0,T;V) \cap L^2(\Omega, A, \tilde{P}; C(0,T;H))$$

$$-du + A(t)u dt = B(t) u d\hat{z} \tag{2.6.7}$$

$$u(x,T) = \Psi(x)$$

$$u \in L^2_Z(0,T;V) \cap L^2(\Omega, A, \tilde{P}; C(0,T;H)) \tag{2.6.8}$$

where $H = L^2(R^n)$; we recall also that $V' = H^{-1}(R^n)$.

If we write (2.6.7), (2.6.8) in integral form, we get an equality between random variables in V'. Note also that

$$L^2_Z(0,T;V) = \{\phi(x,t;\omega) | \text{a.e. t} \quad \phi(.,t;.) \in L^2(\Omega, Z^t; V)$$

and

$$E\int_o^T ||\phi(t)||_V^2 dt < \infty\}.$$

We may consider that $z, \hat{z}$ are R^m valued forward and backward Wiener processes on $\Omega, A, \tilde{P}$, with (2.5.29), and

$$\tilde{E}[z_{t_2} - z_{t_1} | Z^{t_1}] = \tilde{E}[\hat{z}_{t_1} - \hat{z}_{t_2} | \hat{Z}^{t_2}] = 0$$

$$\tilde{E}[(z_{t_2} - z_{t_1})(z_{t_2} - z_{t_1})^* | Z^{t_1}] \tag{2.6.9}$$

$$= \tilde{E}[(z_{t_1} - z_{t_2})(z_{t_1} - z_{t_2})^* | Z^{t_2}]$$

$$= \int_{t_1}^{t_2} D(t) dt \tag{2.6.10}$$

where

$$D(.) \in L^\infty(0,T; L(R^m; R^m)). \tag{2.6.11}$$

We also assume that

$$B(t),\ B^*(t) \in L^\infty(0,T; L(V; H^m)). \tag{2.6.12}$$

2.6.2 Statement of the results

We can state fhe following

Theorem 2.6.1 Assume (2.6.9) to (2.6.12) and

$$2 < Av, v > +\lambda_0 |v|^2 \leq |D^{\frac{1}{2}} B^* v|^2 + \alpha_0 ||v||^2 \tag{2.6.13}$$

$$2 < Av, v > +\lambda_0 |v|^2 \leq |D^{\frac{1}{2}} B v|^2 + \alpha_0 ||v||^2, \tag{2.6.14}$$

$\forall v \in V$, where $<,>$ represents the duality V, V' and $|| \ ||$ represents the norm in V. Then there exists one and only one solution of (2.6.7), (2.6.8).

Proof: In E. Pardoux [7], one uses Galerkin approximation technique. We use here a fixed point argument. We first note that it is sufficient to study

$$dp + (A^* + k)p\, dt = B^* p\, dz \qquad p(0) = p_0.$$

Consider the map $\eta \rightarrow \zeta$ from $L^2_Z(0,T;V)$ into itself (in fact into subspace) defined by the relation

$$d\zeta + (A^* + k)\zeta dt = B^* \eta dz, \quad \zeta(0) = p_0$$

$$\zeta \in L^2_Z(0,T;V) \cap L^2(\Omega, A, \tilde{P}; C(0,T;H)). \tag{2.6.15}$$

We then use the following rule of stochastic calculus in Hilbert spaces:

Let $a(t) \in L^2_Z(0,T;H), b(t) \in L^2_Z(0,T;H^m), \xi_0 \in H$, there exists one and only one process $\xi \in L^2_Z(0,T;V) \cap L^2(\Omega, A, \tilde{P}; C(0,T;H))$ such that

$$\xi(t) + \int_o^t A(s)\xi(s) ds = \xi_0 + \int_o^t a(s) ds + \int_o^t b(s) dz. \tag{2.6.16}$$

and the following energy equality holds

$$|\xi(t)|^2 + 2\int_o^t < A(s)\xi(s), \xi(s) > ds = |\xi_0|^2 + 2\int_o^t (a(s), \xi(s))ds$$
$$+ 2\int_o^t (\xi(s), b(s)).dz + \int_o^t |D^{\frac{1}{2}} b|^2_{Hm} ds. \tag{2.6.17}$$

The equation (2.6.15) enters in the framework of (2.6.16). From the energy equality (2.6.17), we deduce the relation

$$\tilde{E}|\zeta_1(t) - \zeta_2(t)|^2 + 2\tilde{E}\int_o^t < A(\zeta_1 - \zeta_2), \zeta_1, \zeta_2 > ds$$
$$+ 2k\tilde{E}\int_o^t |\zeta_1(s) - \zeta_2(s)|^2 ds = \tilde{E}\int_o^t |D^{\frac{1}{2}} b^*(\eta_1 - \eta_2)|^2 ds \tag{2.6.18}$$

where η_1, η_2 are two elements of $L^2(0,T;V)$ and ζ_1, ζ_2 are the corresponding solutions of (2.6.15).

Now it follows from the assumption (2.6.13) and the equality (2.6.18) that

$$\tilde{E}|\zeta_1(t) - \zeta_2(t)|^2 + 2\tilde{E}\int_o^t < A(\zeta_1 - \zeta_2), \zeta_1, \zeta_2 > ds + 2k\tilde{E}\int_o^t |\zeta_1(s) - \zeta_2(s)|^2 ds$$
$$\leq 2\tilde{E}\int_o^t < A(\eta_1 - \eta_2), \eta_1 - \eta_2 > ds$$
$$-\alpha_0\tilde{E}\int_o^t ||\eta_1 - \eta_2||^2 ds + \lambda_0\tilde{E}\int_o^t |\eta_1 - \eta_2|^2 ds.$$

Define on V a norm

$$|||v|||^2 =< Av, v > + \gamma|v|^2, \gamma > \lambda.$$

By a convenient choice of γ and k, we deduce

$$\tilde{E}\int_o^t |||\zeta_1 - \zeta_2|||^2 dt \leq c_0\tilde{E}\int_o^t |||\eta_1 - \eta_2|||^2 dt, c_0 < 1,$$

hence the map $\eta \to \zeta$ defined by (2.6.15) is a contraction, and the desired result follows.

2.6.3 Interpretation

We can then state the

Theorem 2.6.2 Under the assumptions of Theorem 2.6.1, one has

$$p(T)(\Psi) = \int p(x,T)\Psi(x)dx, \quad \forall\Psi \text{ continuous and bounded} = \int u(x,0)p_0(x)dx. \tag{2.6.19}$$

Proof: It is sufficient to assume Ψ smooth. Consider the equation

$$\frac{\partial v}{\partial t} - A(t)v + i\beta^* Bv = 0$$

$$v(x,T) = \Psi(x)$$

then one has the relation

$$(v(0), p_0) = E\theta_T p(T)(\Psi)$$

$$= \tilde{E}\theta_T(p(T), \Psi) \tag{2.6.20}$$

where θ_t has been defined in (2.1.17).

Define also $\hat{\theta}_t$ by

$$\hat{\theta}_t = -i\hat{\theta}_t \beta_t^* D_t^{-1} d\hat{z}_t, \hat{\theta}_T = 1. \tag{2.6.21}$$

Using backward stochastic calculus, one checks that

$$v(T) = \tilde{E}u(t)\hat{\theta}_t$$

is the solution of

$$\frac{\partial v}{\partial t} - A(t)v + i\beta^* Bv = 0$$

$$v(x,T) = \Psi(x).$$

Therefore, one has

$$v(0) = \tilde{E}u(0)\hat{\theta}_0$$

$$(v(0), p_0) = \tilde{E}(u(0), p_0)\hat{\theta}_0.$$

However,

$$\hat{\theta}_0 = \theta_T$$

and the result follows.

2.6.4 Duality

We have seen in the previous theorem that

$$(u(0), p_0)) = (u(T), p(T)).$$

In fact, one has the following

Theorem 2.6.3 One has

$$(u(t), p(t)) = (u(T), p(T)), \qquad \forall t.$$

Proof: Note that

$$\theta_T = \theta_t \hat{\theta}_t$$

hence

$$\tilde{E}\theta_T(u(t), p(t)) = \tilde{E}(u(t)\hat{\theta}_t, p(t)\theta_t) = (\tilde{E}(u(t)\hat{\theta}_t, \tilde{E}p(t)\theta_t)$$

$$= (v(t), \chi(t))$$

and

$$\frac{\partial \chi}{\partial t} + A^* \chi = i\beta^* B^* \chi$$

$$\chi(0) = p_0.$$

One checks that

$$(v(t), \chi(t)) = (v(T), \chi(T)) = (\Psi, \tilde{E}\theta_T p(T)) = \tilde{E}\theta^T(\Psi, p(T))$$

which is the desired result.

REFERENCES

[1] FUJISAKI, M., KALLIANPUR, G., KUNITA, H., (1972). *Stochastic differential equations for the nonlinear filtering problem.* Osaka J. Math. 9, 1, 19-40.

[2] HAUSSMANN, U., PARDOUX, E., *Time reversal of diffusions.* Publication de Mathématiques Appliquées, Marseille-Toulouse, 84-3, Université de Provence.

[3] KALLIANPUR, G., (1980). *Stochastic Filtering Theory.* Applications of Mathematics, Springer Verlag, Berlin, N.Y.

[4] KUNITA, H., WATANABE, S., (1967). *On square integrable martingales.* Nagoya Math. J. 30, 209-245.

[5] LIPSTER, R. S., SHIRYAEV, A. N., (1977). *Statistics of Random Processes.* Vol. I, II, Springer Verlag, N.Y.

[6] MEYER, P. A., (1966). *Probabilités et potentiel.* Hermann, Paris.

[7] PARDOUX, E., (1979). *Equations du filtrage non linéaire, de la prédiction et du lissage.* Stochastics, 3, 127-168.

[8] PICARD, J., (Dec. 1986). *Non linear filtering of one dimensional diffusions in the case of a high signal to noise ratio.* SIAM Appl. Math. Vol. 46, No. 6, Dec. 1986.

3

AN INTRODUCTION TO MALLIAVIN CALCULUS AND SOME OF ITS APPLICATIONS

Dominique Michel
and
Etienne Pardoux

INTRODUCTION

A new part of the theory of stochastic processes, initiated by P. Malliavin, and developed among others by D. Stroock, J. M. Bismut, S. Watanabe and many others has recently emerged under the name of "the Malliavin Calculus." This is a stochastic calculus of variation which is in particular able to give powerful criterions for the law of a given functional of the Brownian motion to possess a density. One can distinguish three parts in the Malliavin Calculus. First of all, it relies on the use of differential operators which apply to functionals on Wiener space, associated Sobolev spaces, and an integration by parts formula which relates the derivations on Wiener Space and the Itô integral. Second, it contains a criterion, essentially in terms of the "Malliavin covariance matrix," or a similar quantity in Bismut's approach for a random vector defined on Wiener space to possess a density, or even a smooth density. The third circle of ideas concerns the way in which one can give in specific examples sufficient conditions for the above criterion to be satisfied. The most studied example is that of the solution of a stochastic differential equation, for which the Malliavin Calculus produces a probabilistic proof of Hörmander's "sum of squares theorem."

It has become clear during recent years that the Malliavin Calculus is not just a fancy way of making the probabilistic theory of diffusion processes self-contained (i.e. suppressing the need for borrowing results from analysis), but that it is a powerful tool for analyzing Wiener functionals, which is of interest to theoreticians as well as to specialists of stochastic control and nonlinear filtering.

This set of notes, which reflects the contents of a series of lectures given by the authors at the Systems Research Center, University of Maryland at College Park, aims at introducing and motivating the study of the Malliavin Calculus.

Section 3.4.1 explains the ideas of Malliavin and Bismut in the simple case of a finite dimensional probability space. In Section 3.4.2, we present the basic definitions and results of the Malliavin calculus on Wiener space, following Bismut and Zakai. Section 3.4.3 explains how the sufficient condition for the existence of a density follows from Hörmander's condition, in the case of the solution of a stochastic differential equation. In Section 3.4.4, we study the same problem for the conditional law in a filtering problem. Section 3.4.5 presents the application of the Malliavin calculus to stochastic partial differential equations, and how it allows to prove the non existence of finite dimensional statistics in certain nonlinear filtering problems. Finally in Section 3.4.6 we show how the differential calculus on Wiener space can be used to generalize Itô's and Stratonovich's stochastic integrals and differential calculus.

The aim of this text is to state the main results and introduce some of the relevant techniques. No detailed proof is given; only some proofs are sketched. We have included an extended bibliography at the end. Each Section is followed by bibliographical comments.

We want to thank the Systems Research Center and in particular its director John Baras for inviting us to give this series of lectures and for their warm hospitality.

PRELIMINARY

Let $(\Omega, \mathcal{F}, P)$ be a probability space and $\Phi : \Omega \to \mathrm{R}^n$ a measurable map. If P^Φ is the image measure of P under Φ the question is:

(i) does P^Φ have a density w.r.t. Lebesgue measure on R^n?

(ii) is this density regular?

In the case where Ω is finite dimensional, some answers can be given using the differential calculus on Ω. When $\Omega = C_0(\mathrm{R}_+, \mathrm{R}^m)$, the m dimensional Wiener space of continuous paths in R^m starting at 0, the Banach structure on Ω is too strong to give an answer to the problem, essentially because most of the Wiener functionals are not even continuous for this structure. The idea of Malliavin was to define a new notion of regularity of a Wiener functional and to use it to perform an integration by parts on Ω.

The answer to the above question is then given by the following lemma.

Lemma. Let μ be finite Radon measure on R^n. Assume that there exist $m \in \mathrm{N}$, $m \geq 1$ and $C \geq 0$ such that:

$$\forall \alpha, |\alpha| \leq m, \quad \forall \varphi \in C_b^\infty(\mathrm{R}^n), \quad \left| \int \mathrm{D}^\alpha \varphi(\mathrm{x}) \mu(\mathrm{dx}) \right| \leq \mathrm{C} \|\varphi\|_\infty$$

then μ has a density w.r.t. Lebesgue measure and, if $m > n$, this density is in $C_b^k(\mathrm{R}^n)$ with $k = m - n - 1$.

3.1 A FINITE DIMENSIONAL ANALOG OF MALLIAVIN CALCULUS

In this paragraph, $(\Omega, \mathcal{F}, \mathcal{P})$ is $\left(\mathrm{R}^N, \mathrm{B}, \ \mathrm{g}(\mathrm{x})\,\mathrm{dx}\right)$ where B is the Borel field on R^N and g a C^∞ positive function on R^N with integral 1.

$$\Phi \text{ is a } C^\infty \text{ map from } \mathrm{R}^N \text{ to } \mathrm{R}^n$$

A natural assumption for the existence of a density for $P^\Phi = \Phi_*(P)$ is that Φ be a submersion, i.e. that the differential of Φ be of maximal rank, namely n. So, from now on, we assume:

$$\Phi \text{ is a submersion.} \tag{H_1}$$

If Φ is a submersion and is proper (i.e. B is a bounded set of $\mathrm{R}^n \Longrightarrow \Phi^{-1}(\mathrm{B})$ is a bounded set of R^N), it is not difficult to prove that P^Φ has a smooth density. If Φ is not proper, there is a density but nothing can be said about its regularity.

In that case, we look for necessary conditions which ensure that the following integration by parts formula is true:

$$\forall \varphi \in C_b^\infty(\mathrm{R}^n), \quad \forall i,\ 0 < i \leq n, \quad \exists\, B^\varphi \in L^1(\Omega),$$

$$\int D_i\, \varphi(x)\ P^\Phi(dx) = E(D_i\, \varphi \circ\ \Phi) = E(\varphi \circ \Phi \cdot B^\varphi) \tag{3.1.1}$$

From (3.1.1) and Lemma 0, we get the answer.

In the following, we give two methods which lead to this integration by parts formula and which can be generalized to the infinite dimensional case of the Wiener Space.

3.1.1 The Malliavin Method

It is adapted to the particular case where P is the canonical gaussian measure on R^N, i.e. $g(y) = (2\pi)^{-\frac{n}{2}} \exp\left(-\frac{|y|^2}{2}\right)$.

The Ornstein-Uhlenbeck operator

$$L = \Delta - x \cdot \nabla$$

is self adjoint with respect to the gaussian measure i.e. :

$$E(LF\ \cdot\ G) = E(F\ \cdot\ LG), \tag{3.1.2}$$

provided the above quantities are well defined.

Furthermore, let F, G and $F\,G \in\ D(L)$, where $D(L) \triangleq \{\Phi\ \in L^2(\mathrm{R}^\mathrm{N});$ $L\,\Phi\ \in\ L^2(\mathrm{R}^\mathrm{N})\}$, and define:

$$\Gamma(F,\ G) = \frac{1}{2}\ (L(FG)\ -\ FLG\ -\ GLF)$$

If F and G are C^1 functions on R^N, it holds:

$$\Gamma(F,\ G) = \nabla\, F\ \cdot\ \nabla\, G \tag{3.1.3}$$

Formulas (3.1.2) and (3.1.3) are the keys for the integration by parts formula.

Let $J = D\ \Phi$ be the differential of $\Phi\ \cdot\ J$ is an $(N,\ n)$ matrix and assumption (H_1) is equivalent to the invertibility of the $(n,\ n)$ matrix

$$A = J^* J = \left((\nabla\Phi^i\ \cdot\ \nabla\Phi^j)\right)_{i,j=1 \text{ to } n} \tag{H_1}$$

which is called the Malliavin matrix.

The way of getting the integration by parts formula is the following.

$$E(D_i\ \varphi\ \circ\ \Phi) = \sum_{j,k} E\left(D_j\ \ (\varphi\ \circ\ \Phi)\ D_j\ \Phi_k\ A^{-1}_{k\ i}\right)$$

Assume:

$$\varphi \circ \Phi,\ \Phi,\ \Phi\,(\varphi \circ \Phi),\ A^{-1}_{ki},\ A^{-1}_{ki}\ \Phi_k \text{ are in } D\,(L) \tag{H_2}$$

Then

$$\begin{aligned} E\,(D_i\varphi \circ \Phi) &= \sum_k E\left(\nabla\varphi \circ \Phi \cdot \nabla\Phi_k \cdot A^{-1}_{ki}\right) \\ &= \frac{1}{2}\sum_k E\left(\{L\,(\varphi \circ \Phi \cdot \Phi_k) - \varphi \circ \Phi L\Phi_k - \Phi_k L\varphi \circ \Phi\}\ A^{-1}_{ki}\right) \\ &= E\,(\varphi \circ \Phi \cdot B^{\varphi}) \end{aligned}$$

where $B^{\varphi} = \frac{1}{2}\{\Phi_k\ L\ A^{-1}_{ki} - A^{-1}_{ki}\ L\ \Phi_k - L\,(\Phi_k\ A^{-1}_{ki}\}$.

It is possible to get, in a similar way, an integration by parts formula up to any order with allows to prove the existence of a smooth density for P^{Φ}.

Comments on the generalization to Wiener space. The natural way of generalizing the operator L to an infinite dimensional setting is to define it by its spectral decomposition.

If $\alpha = (\alpha_1, \ldots, \alpha_N)\ \in rmN^N$, define:

$$H_\alpha(y) = \prod_{i=1}^{N} H_{\alpha_i}\ (y_i),\ y \in \mathrm{R}^{\mathrm{N}}$$

$$\text{where}\quad H_k(x) = \frac{(-1)^k}{(k!)^{\frac{1}{2}}}\ e^{\frac{x^2}{2}}\ \frac{\partial^k}{\partial x^k}\left(e^{\frac{-x^2}{2}}\right)$$

is the k-th Hermite polynomial on R.

The Hermite polynomials $(H_\alpha)_{\alpha\ \in\ rmN^N}$ are eignevectors of L. More precisely:

$$LH_\alpha = -|\alpha|\ H_\alpha.$$

Furthermore, the set $\{H_\alpha,\ \alpha \in rmN^N\}$ is an orthonormal basis of $L^2(\mathrm{R}^{\mathrm{N}}, \mathrm{P})$. So, the domain of L in $L^2(\mathrm{R}^{\mathrm{N}}, \mathrm{P})$ is:

$$D(L) = \{\Phi \in L^2(\mathrm{R}^{\mathrm{N}}, \mathrm{P}) \ / \ \Phi = \sum_{\alpha \in rmN^N} \lambda_\alpha H_\alpha,$$

$$\sum_{\alpha \in rmN^N} |\alpha|^2 \lambda_\alpha^2 < +\infty\}$$

and if $\Phi \in D(L)$, by definition

$$L\,\Phi = - \sum_{\alpha \in rmN^N} |\alpha| \lambda_\alpha H_\alpha.$$

All this program can be developed in the Wiener space setting: the multiple Wiener integrals stand for the Hermite polynomials, and the fact that they form an orthonormal basis of $L^2(\Omega, P)$ is the Wiener chaos decomposition of a square integrable functional of the brownian motion.

So, if Φ is in $L^2(\Omega, P)$ with its values in R^n, the assumptions for P^Φ to have a smooth density are: (H_1') the Mallavian matrix:

$$A = ((\Gamma(\Phi^i,\ \Phi^j)))_{i,j=1\ to\ n}$$

is invertible.

(H_2) Φ and A^{-1} are "regular" and have good properties of integrability.

This can be rewritten in the following way:

(H_1'') the Malliavin matrix A is invertible and $|\det A|^{-1} \in L^p(\Omega,\ P)\ \forall p \geq 1$.

(H_2') Φ is "regular" and has good properties of integrability as well as its "derivatives."

3.1.2 The Bismut method

The idea of Bismut is not related to the Gaussian character of the measure but to the quasi invariance property of the measure with respect to some transformations on Ω.

If $g = 1$, (of course, P is not a probability measure but, any-way, we can look for an integration by parts formula), P is Lebesgue measure and, so, is invariant under translations.

Take $H : \mathrm{R}^{\mathrm{N}} \longrightarrow \mathrm{R}$ a C^1 map, P integrable. $\int_{\mathrm{R}^{\mathrm{N}}} \varphi \circ \Phi(y+a)\ H(y+a)\,dy$ is indepcndant of a. By differentiating $w.r.t.a$, we get, under some more regularity assumptions,

$$\int_{\mathrm{R}^{\mathrm{N}}} \sum_{j=1}^{n} D_j\, \varphi \circ \Phi(y)\ D_i\ \Phi_j(y)\ H(y)\,dy = -\int_{\mathrm{R}^{\mathrm{N}}} \varphi \circ \Phi(y)\ D_i\ H(y)\,dy \quad (3.1.4)$$

Under assumption (H_1'), define:

$$H_{ik}(y) = \sum_{j=1}^{N} D_j\ \Phi_i(y)\ A_{jk}^{-1}(y)$$

Write (3.1.4) with $H = H_{ik}$ and sum over i. We get:

$$\int_{\mathrm{R}^{\mathrm{N}}} D_k\, \varphi \circ \Phi(y)\,dy = -\int_{\mathrm{R}^{\mathrm{N}}} \varphi \circ \Phi(y) \sum_{i,j=1}^{N} D_i \left(D_j\ \Phi_i\ A_{jk}^{-1}\right)(y)\,dy$$

This is a good formula of integration by parts if:

$$\sum_{i,j=1}^{N} D_i \left(D_j\ \Phi_i\ A_{jk}^{-1}\right) \in L^1\left(\mathrm{R}^{\mathrm{N}}, \mathrm{dx}\right) \quad (H_3)$$

When $g \neq 1$, the measure P has a priori no invariance property but it is quasi invariant under any C^1 differomorphism Ψ (i.e. $\Psi_*\ P$ is absolutely continuous $w.r.t.\ P$) because of the change of variables formula:

$$dP^{\Psi}(y) = \frac{g\left(\Psi^{-1}(y)\right)}{g(y)}\ |\det\ \Psi|^{-1}(y)\ dP(y)$$

So, the integral is:

$$I(\Psi) = \int_{\mathrm{R}^{\mathrm{N}}} \varphi \circ \Phi \circ \Psi(y)\ dP^{\Psi}(y)$$

In order to differentiate $w.r.t\ \Psi$, we choose a one parameter family of diffeomorphisms, for example the flow associated to a vector field X an R^{N} i.e $\Psi_t(y)$ is the solution at time t of the equation:

$$\begin{cases} dy_t = X(y_t) \\ y_0 = y \end{cases}$$

Now, we write: $\frac{d}{dk}\, I(\Psi_t)_{t=0} = 0$:

$$\int_{\mathrm{R}^{\mathrm{N}}} X(\varphi \circ \Phi)(y)\,dP(y) = -\int_{\mathrm{R}^{\mathrm{N}}} \varphi \circ \Phi(y)\ \frac{\operatorname{div} gX}{g}(y)\,d\ P(y) \quad (3.1.5)$$

$\frac{\text{div } gX}{g}$ is the divergence of the vector field X under the measure P.

In a similar way that we have chosen H in the Malliavin method, we have to choose X in order to get the integration by parts formula.

Since

$$X(\varphi \circ \Phi) = \sum_{i=1}^{N} \sum_{j=1}^{n} X_i \, D_j \, \varphi \circ \Phi \, D_i \, \Phi_j$$
$$= \sum_{j=1}^{n} D_j \, \varphi \circ \Phi \, (X \cdot \nabla \Phi_j), \tag{3.1.6}$$

we need the following assumption:

(H_4) There exist vector fields $Y_1, \ldots, Y_n$ such that the matrix $((Y_i \,.\, \nabla \, \Phi_j))$ is invertible.

Actually, if (H_4) is valid, let B be the inverse matrix, and define:

$$X^i = \sum_{j=1}^{n} B_{i\,j} \, Y_j$$

Then: $X^i(\varphi \circ \Phi) = D_i \varphi \circ \Phi$

So, if

$$\forall i = 1 \ldots n, \qquad \frac{\text{div } gX_i}{g} \in L^1(\Omega, \, P).$$

the integration by parts formula holds.

In the finite dimensional setting, it is easy to show that (H_1') and (H_4) are equivalent. In the infinite dimensional setting, this is no longer the case.

Comments about the generalization. As we are only interested in the behaviour of the flow Ψ_t near $t = 0$, it is equivalent and easier to take for Ψ_t the linearized flow $y + t\, X(y)$. When Ω is the Wiener Space, we need transformations of $\Omega, \omega \longrightarrow \omega + t\, X(\omega)$, leaving the Wiener measure quasi invariant. Some of them are given by Girsanov theorem.

Girsanov's Theorem. Let u: $[0,\ 1] \times W_1 \longrightarrow \mathrm{R}$ be an adapted process such that,

$$E\left(\exp\left(\lambda \int_0^1 u_s^2 \, ds\right)\right) < \infty \quad \text{for a } \lambda > \frac{1}{2}.$$

Then the law of the process $\omega. + \int_0^{\cdot} u_s \, ds$ is absolutely continuous with respect to Wiener measure, the density being:

$$exp\left(-\int_0^1 u_s \, d\omega_s - \frac{1}{2}\int_0^1 u_s^2 \, ds\right)$$

By means of this theorem, it is possible to get an integration by parts formula analog to (3.1.5) on Wiener Space. By the way, it is necesary to define the analog $X \cdot \nabla \Phi$ when Φ is a Wiener satisfies the assumptions of Girsanov theorem.

3.1.3 Bibliographical comments

This section was inspired by similar expositions in Bismut [$A1$] and Stroock [$A13$].

3.2 THE BISMUT - ZAKAI METHOD FOR MALLIAVIN CALCULUS

Let Ω denote the space $C(\mathrm{R}_+, \mathrm{R}^m)$ equipped with the topology of uniform convergence on compact sets, $\mathcal{F}$ the Borel σ-field on Ω, P the standard Wiener measure, and let $\{W_t(\omega) = \omega(t), t \geq 0\}$.

3.2.1 The polynomial functionals

Definition. A polynomial Wiener functional is a map $\Phi : \Omega \longrightarrow \mathrm{R}$ defined by:

$$\Phi = f\left(\delta_{i_1}(h_1), ., \delta_{i_p}(h_p)\right)$$

Where:

(i) f is a polynomial function on R^p

(ii) $\{h_1, \ldots, h_p\}$ is an orthonormal set of $L^2(\mathrm{R}_+)$

(iii) $\delta_i(h) = \int_0^{+\infty} h(s) \, dW^i(s)$

$\mathcal{P}(\Omega)$ is the set of polynomial Wiener functionals.

Proposition. The set of polynomial functionals on Ω is dense in $L^2(\Omega, P)$.

3.2.2 The directional derivative of a Wiener functional

If $\Phi(\omega) = f\left(\delta_{i_1}(h_1), \ldots, \delta_{i_p}(h_p)\right)$ is a polynomial functional and the h_i are C^1 functions with compact support, then:

$$\delta_i(h_j) = -\int_0^{+\infty} W_s^i \, \dot{h}_s^j \, ds$$

and Φ is Frechet differentiable i.e. , if $g \in L^2(\mathrm{R}+;\ \mathrm{R}^m)$

$$\begin{aligned} D_G\,\Phi &= \lim_{\varepsilon\to 0} \frac{\Phi\left(\omega + \varepsilon \int_0^{\cdot} g(s)\,ds\right) - \Phi(\omega)}{\varepsilon} \\ &= \sum_{j=1}^{p} \sum_{\{k;\ i_k=j\}} \frac{\partial f}{\partial x_k}\,(\delta(h))\,(h_k,\ g_j) \end{aligned}$$

where (,) denotes the scalar product in $L^2(\mathrm{R}_+;\ \mathrm{R}^m)$ and $G_s = \int_0^s g(u)\,du$. This motivates the following definition:

Definition. If Φ is a polynomial functional on Ω and $g \in L^2(\mathrm{R}_+;\ \mathrm{R}^m)$, define:

$$D_G\,\Phi\,(\omega) = \sum_{j=1}^{p} \sum_{\{k;\ i_k=j\}} \frac{\partial f}{\partial x_k}\,(\delta(h))\,(h_k,\ g_j)$$

$$\text{where}\quad G_s = \int_0^s g(u)\,du.$$

Notation. Denote by $(D_t\,\Phi)_{t\in\mathrm{R}^+}$ the element of $L^2(\mathrm{R}^+\times\Omega,\ \mathrm{R}^m)$ such that: $D_G\,\Phi = (D\cdot\Phi, g)$ i.e. $D_t^j\Phi = \sum_{\{k;\ i_k=j\}} \frac{\partial f}{\partial x_k}\,(\delta(h))\,h_k\,(t)$

On $\mathcal{P}(\Omega)$, we define the following norm:

$$\begin{aligned} \|\Phi\|_{2,\,1} &= \left\{E\left(\Phi^2 + \sum_{i=1}^{m} \int_o^{+\infty} (D_t^i\,\Phi)^2\,dt\right)\right\}^{\frac{1}{2}} \\ &= \left[E\left(f^2(\delta(h)) + \|\nabla f\|^2(\delta(h))\right)\right]^{\frac{1}{2}} \end{aligned}$$

Definition. $H^{2,1}$ is the closure of $\mathcal{P}(\Omega)$ in $L^2(\Omega)$ for the norm $\|\ \|_{2,1}$

It can be shown that the gradient operator D extends to $H^{2,1}$. The extension is called the derivative on Wiener Space.

3.2.3 The integration by parts formula

Theorem. If Φ is in $H^{2,1}$ and $u \in L^2((0,\ t)\times\Omega,\ \mathrm{R}^m)$ is adapted, then:

$$E\left(\sum_{i=1}^{m} \int_0^t D_s^i\,\Phi\,u_s^i\,ds\right) = E\left(\Phi \sum_{i=1}^{m} \int_0^t u_s^i\,dw_s^i\right)$$

This is an extension to Wiener space of the formula (5) of Section one.

Remark. The gradient operator D can be thought of as a linear continuous operator from $H^{2,1} \subset L^2(\Omega)$ into $L^2(\mathrm{R}_+\times\Omega;\ \mathrm{R}^m)$. Clearly, D has an adjoint D^* which maps

$L^2(\mathrm{R}_+ \times \Omega; \mathrm{R}^m)$ into $(H^{2,1})'$, where $(H^{2,1})'$, the dual space of $H^{2,1}$, is a space of "distributions over Wiener space." Let us now define the operator δ as follows.

$$\text{Dom } \delta = \{u \in L^2(\mathrm{R}_+ \times \Omega;\ \mathrm{R}^m);\ \mathrm{D^*u} \in \mathrm{L}^2(\Omega)\}$$

$$\text{For } u \in \text{Dom}(\delta), \quad \delta\, u \stackrel{\Delta}{=} D^* u$$

It follows from Riesz's representation theorem that an equivalent definition of δ is as follows.

Definition. Dom $\delta = \{u \in L^2(\mathrm{R}_+ \times \Omega, \mathrm{R}^m)$, such that there exists a constant c with: $|E(D \cdot \Phi, u)| \leq c\|\Phi\|_2,\ \forall\, \Phi \in H^{2,1}\}$.

For $u \in$ Dom $\delta, \delta u$ is the unique random variable which satisfies:

$$E(D\ .\ \Phi,\ u) = E(\Phi\ \delta u) \qquad \forall\, \Phi \in H^{2,1}$$

The operator δ is called the Skorohod integral. It will be studied in § 6 below.

3.2.4.1 Theorem

Let Φ be in $L^2\,(\Omega)$. Assume that Φ is in $H^{2,1}$ and that there exists $u \in$ Dom δ such that:

(i) $(D \cdot \Phi,\ u) \in H^{2,\,1}$

(ii) $(D \cdot \Phi, u) > 0$ a.s.

Then P^Φ has a density.

Proof: Let $\varphi \in C_b^\infty(\mathrm{R})$. Then $\varphi \circ \Phi \in H^{2,1}$.

Furthermore, if $\Psi = \dfrac{1}{\varepsilon + (D\ .\ \Phi,\ u)}$, then $\varphi \circ \Phi\ .\ \Psi$ is in $H^{2,\,1}$ by assumption. Then, the following integration by parts holds:

$$E\left((D\,(\varphi \circ \Phi \cdot \Psi),\ u)\right) = E\left((\varphi \circ \Phi)\,\Psi \delta u\right)$$

$$\Longleftrightarrow E\,(\varphi' \circ \Phi(D \cdot \Phi,\ u)\,\Psi) = E\left[(\varphi \circ \Phi)(\Psi\ \delta u) - (D \cdot \Psi,\ u)\right]$$

So, if P_ϵ is the probability law on Ω such that $\dfrac{dP_\epsilon}{dP} = \dfrac{(D \cdot \Phi, u)}{\varepsilon + (D \cdot \Phi, u)}$, then P_ϵ^Φ possesses a density. Now, if B is a borel set of Lebesgue measure zero: $P_\epsilon^\Phi\ (B) = P_\epsilon\,(\Phi \in B) = 0$. By the monotone convergence theorem, $P\,(\Phi \in B) = 0$. So, P^Φ is absolutely continuous w.r.t Lebesgue measure.

Further integrations by parts lead to the smoothness of the density. Let us now state the multidimensional version of Theorem 3.2.4.1.

3.2.4.2 Theorem

Let $\Phi = (\Phi_1, \ldots, \Phi_d)' \in (L^2(\Omega))^d$. Assume that Φ_i is in $H^{2,1}, i = 1, \ldots, d$, and that there exists $u_1, \ldots, u_d \in \text{Dom } \delta$ such that:

(i) $(D \cdot \Phi_i,\ u_j) \in H^{2,1}; 1 \leq i, j \leq d$

(ii) $(D \cdot \Phi_i,\ u_j)_{1 \leq i,j \leq d} > 0$ a.s.

Then P^{Φ} has a density with respect to Lebesgue measure on R^n.

3.2.5 Bibliographical comments

The abstract presentation of Bismut's methodology (independently of the particular application to SDE's) can be found (in the case of one- dimensional processes) in Zakai [$A17$], from which Theorem 2.4.1 is borrowed. For a systematic exposition of the theory of Sobolev spaces over Wiener space, we refer to Watanabe [$A15$].

3.3 THE APPLICATION TO STOCHASTIC DIFFERENTIAL EQUATIONS

In this paragraph, we study the particular case where Φ is the solution at time t of a stochastic differential equation, i.e.

$$dx_t = X_0\,(x_t)\ dt + \sum_{i=1}^{m} X_i\,(x_t)\ \circ dw_t^i \tag{3.3.1}$$

$X_0,\ X_1, \ldots,\ X_m$ are C_b^{∞} maps from R^n to R^n that we consider as vector fields on R^n in the sense that we identify X_i with the first order partial differential operator $\sum_{j=1}^{n} X_i^j \frac{\partial}{\partial x_j}$. Equation (1) is understood in the sense of Stratonovich.

It is natural to associate to (1):

its infinitesimal generator: (3.3.2)

$$L = X_0 + \frac{1}{2} \sum_{i=1}^{m} X_i^2$$

a controlled ordinary differential equation: (3.3.3)

$$\dot{X}_t^u = X_0(x_t^u) + \sum_{i=1}^{m} X_i(x_t^u)\,\dot{u}_t^i$$

$$x_0^u = x_0$$

$$\text{where } u \in H^1(\mathrm{R}, \mathrm{R}^m)$$

The links between (3.1), (3.2) and (3.3) are the following:

3.3.1 Theorem

a. If μ_t is the law of x_t, solution of (3.3.1), then μ_t is the solution of the following PDE written in the distributional sense:

$$\frac{d\mu_t}{dt} - L^*\ \mu_t = 0$$

where L^* is the adjoint of L in the L^2 sense.

b. Support theorem:

$$\text{If } A(t, x_0) = \{X_t^u,\ u \in H^1\} \text{ and}$$

$$A(x_0) = \cup_{t>0} A(t,\ x_0),\ \text{then :}$$

$$\text{support } \mu_t = \overline{A(t,\ x_0)}$$

$$\text{support}\left(\int_0^\infty e^{-\alpha t}\ \mu_t\ dt\right) = \overline{A(x_0)}.$$

So, the existence of a smooth density for μ_t is in relation with:

1. the hypoellipticity of the operation $\frac{\partial}{\partial t} - L^*$. Recall that a partial differential operator A is said to be hypoelliptic if:

   $$A\varphi = \Psi \text{ on } U$$

   $$\Psi_{|U} \in C^\infty(U) \Longrightarrow \varphi_{|U} \in C^\infty(U)$$

   U being an open set of R^n.
2. the accessibility property for the controlled ODE (3.3.3).

The assumptions needed for solving this problem are in terms of the following Lie algebras of vector fields:

1. $\mathcal{A} = LA(X_0,\ X_1, \ldots, X_m)$
2. $\mathcal{B} = LA(X_1, \ldots, X_m)$
3. $\mathcal{J}$ = ideal generated by $\mathcal{B}$ in $\mathcal{A}$

The main results concerning these problems are presented in the following picture:

3.3.2 Theorem

$\dim \mathcal{A}(x) = n, \ \forall x \in \mathrm{R^n}$	$\dim \mathcal{J}(x) = n, \ \forall x \in \mathrm{R^n}$	$\dim \mathcal{B}(x) = n, \ \forall x \in \mathrm{R^n}$
L is hypoelliptic	$\frac{\partial}{\partial t} - L^*$ is hypoelliptic	$\frac{\partial}{\partial t} - L^*$ is hypoelliptic
$\int_0^{+\infty} e^{-\alpha t} \mu_t \, dt$ has a smooth density	μ_t has a smooth density.	μ_t has a smooth density, everywhere strictly positive.
$\mathring{A}(x) \neq \emptyset, \ \forall x \in \mathrm{R^n}$	$\mathring{A}(t,x) \neq \emptyset, \ \forall x \in \mathrm{R^n}$ $t > 0$	$A(t,x) = \mathrm{R^n}$, $\forall x \in \mathrm{R^n}$ $t > 0$

3.3.3 Comments

1. The reason why the Lie algebras occur in the accessibility problem is essentially the following: if the vector fields X_i commute, for a control u with piecewise constant derivative, x_t^u is a composition of the flows associated to each X_i. If they don't, the flows associated to the brackets occur because of the parallelogram law: follow the vector field X_i during time ε, then X_j during the same time, then $-X_i$ and $-X_j$; up to the second order, it is as if you had followed the field $[X_i,\ X_j]$ during time 4ε, up to a term of order ε^2.

2. The implication $\{\dim A(x) = n, \forall x \in \mathrm{R}^n \Longrightarrow \mathrm{L}$ is hypoelliptic$\}$ is Hörmander's theorem. To obtain the implication $\left\{\dim \mathcal{J}(x) = n, \forall\, x \in \mathrm{R}^n \Longrightarrow \frac{\partial}{\partial t} - \mathrm{L}^* \text{ is hypoelliptic}\right\}$, it suffices to apply Hörmander's theorem to the operator on R^{n+1} $\frac{\partial}{\partial t} - \mathrm{L}^*$: then occurs the Lie algebra $\mathcal{C} = LA\left(\frac{\partial}{\partial t} + \widetilde{X}_0, X_i \ldots X_n\right)$ where $\widetilde{X}_0$ is the vector field defined by:

$$\frac{\partial}{\partial t} - L^* = \frac{\partial}{\partial t} + \widetilde{X}_0 - \frac{1}{2}\ \Sigma X_i^2 + c$$

c being a zero degree operator is a function. Actually, $\widetilde{X}_0 = -X_0 + \sum_{i=1}^m \lambda_i X_i$, and the two following conditions are equivalent:

a. $\dim\ \mathcal{C}\,(t,\ x) = n+1 \quad \forall (t,\ x)\ \in \mathrm{R}^{n+1}$

b. $\dim\ \mathcal{J}\,(x) = n \quad \forall_x \in \mathrm{R}^n$

3. Beware that the fact that the support of μ_t is the whole space does not imply that μ_t has a smooth density. The following example is in R^2.

$$X_1 = \frac{\partial}{\partial x}$$
$$X_2 = \frac{\partial}{\partial x} + \varphi\,(x)\ \frac{\partial}{\partial y}$$

where $\varphi : \mathrm{R} \longrightarrow \mathrm{R}$ is a smooth function whose value is 0 on $\mathrm{R}-$.

$$\dim \quad \mathcal{B}\,(x,\ y) = 1 \qquad \text{if } \mathrm{x} \le 0$$

$$\dim \quad \mathcal{B}\,(x,\ y) = 2 \qquad \text{if } \mathrm{x} > 0$$

So, on $\mathrm{R}^- \times \mathrm{R}$, the system $(X_1,\ X_2)$ does not satisfy the range condition. Nevertheless:

$$A\,(t,\ x,\ y) = \mathrm{R}^2, \qquad \forall \mathrm{t}\ \in \mathrm{R}^+.$$

and certainly μ_t does not have a density when $x_0 < 0$. But, we shall see, using Malliavin's calculus, that μ_t has a smooth density when $x_0 > 0$ because the rank condition is needed only at the starting point.

Stroock showed that, if the rank condition fails on a submanifold N of codimension $n-1$, under some technical assumptions on the degeneracy, then μ_t has a smooth density even if the starting point is on N.

3.3.4 The Malliavin, Bismut method

Theorem. Let x_t be the solution of the SDE (1) at time t and μ_t its law. Then, if:

$$\dim \mathcal{J}(x_0) = n,$$

μ_t has a smooth density.

Proof: The proof proceeds in three steps.

First step: $x_t \in H^{2,1}$

This is done by solving (1) by the Picard iteration method, using the rules of calculus for the gradient and passing to the limit. The gradient $D_s x_t$ is a solution of the following SDE:

$$D^i_x x_t = X_i(x_s) + \int_s^t X'_0(x_r)\; D^i_s x_r\; dr + \int_s^t X'_j(x_r)\; D^i_s x_r \circ dw^j_r$$

Let Φ_t be the $n \times n$ matrix valued process solution of:

$$\Phi_t = I + \int_0^t X'_0(x_s)\; \Phi_s\; ds + \int_0^t X'_j(x_s)\; \Phi_s \circ dw^i_s.$$

Remark that Φ_t is nothing else but the jacobian matrix of x_t w.r.t. x_0.

$$\text{Then:}\quad D^i_s x_t = \Phi_t\, \Phi_s^{-1}\, X_i(x_s)$$

Second step: Choose u such that $((Dx^i_t,\, u_j))_{i,j} > 0$. Look at the bilinear form associated to $((Dx^i_t,\, u_j))_{i,j}$,

$$((D_{u_j} x^i_t) \cdot p,\, q) = \sum_{i,\, j} Du_j\; x^i_t\; p_j\; q_i$$

$$= \int_0^t \Phi^j_t\; \Phi_s^{-1}\; X_r\; u^k_j\; p_j\; q_i\; ds$$

$$= \int_0^t \sum_k \left(\Phi_t\, \Phi_s^{-1}\, X_k, q\right) \left(u^k,\, p\right) ds$$

We get the best rank if span $((u^1)', \ldots, (u^m)') = \text{span}(\Phi_t\Phi_s^{-1}X_1, \ldots, \Phi_t\Phi_s^{-1}X_m)$ and, so in particular, if $(u^k)' = \Phi_t\ \Phi_s^{-1}X_k$, where $(u^k)'$ is the column vector whose j-th component is u_j^k.

In that case, the Bismut approach leads to the same condition as the Malliavin approach i.e. to check the invertibility of the Malliavin covariance matrix $A_t = ((Dx_t^i, Dx_t^j))_{i,j}$.

Third step: The invertibility of A_t

Remark first that:

$A_t = \Phi_t C_t \Phi_t^{-1}$

with $C_t = \int_0^t \sum_k (\Phi_s^{-1}X_k)\ (\Phi_s^{-1}X_k)^t\,(x_s)\,ds$

From the invertibility of Φ_t and its boundedness properties, it is clear that it is equivalent to prove $E\left((\det\ A_t)^{-p}\right) < +\infty$ or $E\left((\det\ C_t)^{-p}\right) < +\infty$. We first consider the elliptic case and then the hypoelliptic case.

3.3.4.1 Theorem.

Assume $X_1 \ldots X_m$ generate $\mathrm{R^n}$ at x_0 then C_t is positive definite a.s. and so μ_t has a density.

Proof: For each $q \in \mathrm{R^n}$, $\sum_{\mathrm{k=1}}^{\mathrm{m}} (\mathrm{X_k}\,(\mathrm{x_0})\,, \mathrm{q})^2$ is strictly positive. So, by continuity, $(C_t\ q, q)$ is strictly positive a.s.

3.3.4.2 Theorem.

Assume $\mathcal{J}\,(x_0) = \mathrm{R^n}$. Then C_t is positive definite a.s. and so μ_t has a density

Proof: Define $U_s = \text{span}\left\{\Phi_s^{-1}\ X_i,\ 1 \le i \le m\right\}$, $V_t = \ \cup\ U_s$ and $V_t^+ = \ \cap\ V_s$. By the 3.0.1. law, V_0^+ is a non random space a.s. Assume $V_0^+ \neq \mathrm{R^n}$ and take q in $(V_0^+)^\perp$. Define $\tau = \inf\{s > 0,\ V_s \neq V_0^+\}$. It is clear that, a.s., $\tau > 0$. Then, for $t < \tau$:

$$(q,\ \Phi_t^{-1}X_i) = 0, \quad \forall\, i = 1, \ldots, m.$$

By Itô's formula it holds:

$$(\Phi_t^{-1}X_i, q) = (X_i, q) + \int_0^t (\Phi_s^{-1}\,[X_0, X_i]\,, q)ds$$
$$+ \int_0^t (\Phi_s^{-1}\,[X_j, X_i]\,, q) \circ dW^j$$

Annihilating the martingale part, we get:

$$(q,\ \Phi_t^{-1}[X_j, X_i]) = 0, \quad \forall t < \tau.$$

and by iteration: $(q, \Phi_t^{-1}[X_k, [X_j, X_i]]) = 0$. Then, annihilating the bounded variation part, it comes, $(q, \Phi_t^{-1}[X_0, X_i]) = 0$

Recursively, we can prove that q is orthogonal to every bracket which contradicts the assumption. So, $V_0^+ = \mathrm{R}^n$, and this implies that each V_s is R^n.

The problem of the smoothness of the density is more complicated because we need to prove that $E\left((\det A_t)^{-p}\right) < +\infty$. But, the first idea is the same, namely to give a kind of Taylor expansion of $\Phi_t^{-1} X_i$ by means of the brackets. Before looking at other properties of the density we give a classical example of a hypoelliptic diffusion.

3.3.5 Example: the Heisenberg group

The Heisenberg group H_3 is $\mathrm{C} \times \mathrm{R}$ together with the group law:

$$(z,t)\ (z',t') = (z + \bar{z}',\ t + t' + 2\ \mathrm{Im}\ z\bar{z}').$$

X, Y, T are the left invariant vector fields on H_3 defined by:

$$X = \frac{\partial}{\partial x} + 2y\frac{\partial}{\partial t}, \quad Y = \frac{\partial}{\partial y} - 2x\frac{\partial}{\partial t}, \quad T = \frac{\partial}{\partial t}.$$

They form a basis of the tangent space. Let ξ be the diffusion on H_3 associated to the Kohn laplacian: $\Delta_K = X^2 + Y^2$. Because, $[X,Y] = 4T$, ξ is a hypoelliptic diffusion. It is a solution of the system:

$$d\xi = \begin{pmatrix} 1 \\ 0 \\ 2\xi_2 \end{pmatrix} dw_1 + \begin{pmatrix} 0 \\ 1 \\ -2\xi_1 \end{pmatrix} dw_2$$

So, if $\xi_0 = 0$, we get,

$$\xi = \begin{pmatrix} w_1(t) \\ w_2(t) \\ 2\left(\int_0^t w_2\, dw_1 - \int_0^t w_2 dw_1\right) \end{pmatrix}.$$

The last component is 4 times the area swipped out by the vector $(w_1,\ w_2)$. Paul Levy gave a formula for the density of the law of this area knowing $(w_1,\ w_2)$.

3.3.6 Further properties of the density

3.3.6.1 Positivity Theorem.

If $\mathcal{B}(x) = \mathrm{R}^n$, $\forall x \in \mathrm{R}^n$, then the density of μ_t, $p(t, x)$, is strictly positive for all $(t, x) \in \mathrm{R}_+^* \times \mathrm{R}^n$.

Proof: We use the Feynmann-Kac formula and a time reversal. Let $\alpha \in (0, t)$ and define

$$q(s, x) = p(t-s, x), \quad \text{for } s \leq t - \alpha$$

q is a solution of the following PDE:

$$\left\{ \begin{array}{ll} \dfrac{\partial q}{\partial s} + L^* q = 0, & s \leq t - \alpha \\ q\,(t-\alpha) \text{ given.} & \end{array} \right\}.$$

where $L^* = \frac{1}{2} \sum_1^m X_i^2 - \widetilde{X}_0 - c$ as we saw before. Let y_s^x be the diffusion associated to the operator $\frac{1}{2} \sum_1^m X_i^2 - \widetilde{X}_0$, starting at x.

By the Feynmann-Kac formula, it holds:

$$p\,(t,x) = q\,(0,x) = E\left(q\left(t-\alpha, y_{t-\alpha}^x\right) \exp\left(\int_o^{t-\alpha} c\,(y_s^x)\, ds \right) \right)$$

Since $x \rightarrow q\,(t-\alpha, x) = p\,(\alpha, x)$ is the smooth density of a probability measure, there exists a ball $B \subset \mathrm{R}^n$ and a strictly positive constant $k \in \mathrm{R}$, such that:

$$q\,(t-\alpha, x) \geq k, \quad \forall x \in B$$

Moreover, as $|c|$ is bounded, there exists $\bar{c} \in \mathrm{R}$, such that, $c\,(x) \geq \bar{c}$, $\forall x \in \mathrm{R}^n$. We get then:

$$p\,(t,x) \geq k \exp\left(\bar{c}\,(t-\alpha)\right)\; P\left(y_{t-\alpha}^x \in B\right)$$

Because $\mathcal{B}\,(x) = \mathrm{R}^n$, $\forall x \in \mathrm{R}^n$, the support of the law of $y_{t-\alpha}^x$ is R^n. This implies that:

$$P\left(y_{t-\alpha}^x \in B\right) > 0.$$

Remark. The statement of the theorem can be improved as follows.

Let $t > 0$. Suppose that for some $s \in (0,t)$, $A\,(s,x) = \mathrm{R}^n$ for all $x \in \mathrm{R}^n$. Then $p(t, x) > 0$ for all $x \in \mathrm{R}^n$.

Indeed, the hypothesis has been used only to insure that for some $\alpha \in (0,t)$ and all $x \in \mathrm{R}^n$, the support of the law of $y^x_{t-\alpha}$ is R^n. That support coincides with $\bar{B}(t-\alpha,x)$, where $B(t,x)$ is the accessible set at time t, starting from x, for the control system:

$$\dot{y}^u_t = -X_0(y^u_t) + \sum_{i=1}^m a_i(y^u_t) X_i(y^u_t) + \\ + \sum_{i=1}^m X_i(y^u_t)\dot{u}^i_t$$

which is the same as that associated to:

$$\dot{y}^u_t = -X_0(y^u_t) + \sum_{i=1}^n X_i(y^u_t)\dot{u}^i_t$$

It finally follows easily from a time reversal argument that $A(s,x) = \mathrm{R}^n$ for all $x \in \mathrm{R}^n$ if and only if $B(s, x) = \mathrm{R}^n$ for all $x \in \mathrm{R}^n$.

3.3.6.2 The regularity with respect to the starting point

Let $p(t,x_0,x)$ be the density of μ_t, x_0 being the starting point of the diffusion. As a function of (t,x_0), p satisfies the following backward equation:

$$\frac{\partial p}{\partial t} + Lp = 0.$$

Then, if $\Im(x) = \mathrm{R}^n$, $\forall \mathrm{x} \in \mathrm{R}^n$, $\mathrm{p}(\mathrm{t},\mathrm{x}_0,\mathrm{x})$ is smooth w.r.t. x_0, by Hörmander's theorem. One way of recovering this result by using the Malliavin calculus is to proceed as in the previous paragraph i.e. if φ is a smooth function with compact support:

$$\int p(t,x_0,x)\varphi(x_0)\,dx_0 = E\left(\varphi(y^x_t)\exp\left(\int_0^t c(y^x_s)\,ds\right)\right)$$

by Feynmann-Kac's formula.

Malliavin's calculus applied on the right hand side allows one to get estimates of the type:

$$|\int p(t,x_0,x)\,D^\alpha\varphi(x_0)\,dx_0| \leq c_\alpha\|\varphi\|$$

and then to conclude.

3.3.6.3 The hypoellipticity of L

If $\dim \Im(x) = n, \forall x \in \mathrm{R}^n, \mu_t$ has a density which is smooth with respect to the forward and backward variables. Using localization and integration w.r.t. x_0 Stroock showed that L is hypoelliptic and, so, recovered the result of Hörmander's theorem.

Assume now the more general condition, $\dim \mathcal{A}(x) = n \quad \forall x \in \mathrm{R}^n$.

Stroock's method. It uses the following trick to go back to the previous case. Let W^{m+1} be a Wiener process independent of $W^1, \dots, W^m$, and define:

$$\tilde{x}_t = \begin{pmatrix} x \int_0^t \varrho(W^m(s))ds \\ W^{m+1}(t) \end{pmatrix}$$

where ϱ is a C_b^∞ strictly positive function. Then $\tilde{x}_t$ is a solution of the following SDE:

$$d\tilde{x}_t = \widetilde{X}_0(\tilde{x}_t)\,dt + \sum_1^m \widetilde{X}_k(\tilde{x}_t) \circ dW_t^k + \widetilde{X}_{m+1}(\tilde{x}_t) \circ dW_t^{m+1}$$

with:

$$\widetilde{X}_0 = \begin{pmatrix} \rho X_0 \\ 0 \end{pmatrix}, \quad \widetilde{X}_k = \begin{pmatrix} \sqrt{\rho} X_k \\ 0 \end{pmatrix}, \quad \widetilde{X}_{m+1} = \begin{pmatrix} 0 \\ 1 \end{pmatrix}$$

and $\Im\left(\widetilde{X}_0, \widetilde{X}_1, \dots, \widetilde{X}_{m+1}\right) = \mathrm{R}^{m+1}$ at each point. So, $\widetilde{L} = \frac{1}{2} \sum_1^{m+1} \widetilde{X}_i^2 + \widetilde{X}_0$ is hypoelliptic. As $\widetilde{L}_{|\mathrm{R}^m} = \rho(x_{m+1})L$, the same property holds for L.

Bismut's method. This is a geometric method. We shall only give an idea of the method in the case where moreover:

$$\dim \Im(x) = n - 1, \ \forall x \in \mathrm{R}^n$$

By Frobenius theorem, R^n is a disjoint union of maximal integral submanifolds of $\Im$. The idea of Bismut is to write the flow associated to x_t as a composition of a diffusion flow living in an integral submanifold and a deterministic transverse flow.

More explicitly, let φ_t be the flow associated to X_0. Then:

Proposition. $x_t = \varphi_t(X_t)$ where,

$$dX_t = \varphi_t^{*-1}\, X_i(X_t) \circ dW_t^i$$

Denote by M_{x_o} the integral submanifold of $\Im$ which contains x_o. Then:

Proposition. $X_t(x_0) \in M_{x_0}$, $\forall t \in \mathrm{R}^+$ and the law of X_t has a density on C M_{x_0}.

The hypoellipticity of L is obtained by integration along the flow φ_t.

3.3.7 Bibliographical comments

The equation in Theorem 3.1.a is often called the forward equation, or the Fokker-Planck equation. It can be found in most standard textbooks on stochastic differential equations and diffusion processes. Theorem 3.1.b. is due to Stroock-Varadhan [A 15].

The first two lines of Theorem 3.2 is Hörmander's sum of squares theorem, whose probabilistic proof is essentially the subject of the rest of the section. The third line of Theorem 3.2 is borrowed from results in control theory, see e.g. A. Isidori: Nonlinear Control Systems, Lecture Notes in Control and Info. Sci. 72, Springer Verlag (1985) or C. Lobry: Bases mathématiques de la théorie des systèmes asservis non linéaires, mimeographed, Univ. de Bordeaux (1976).

Bismut's version of the Malliavin calculus applied to SDEs appears in [A1]. In his paper, Bismut makes a heavy use of the theory of flows. This may obscur the exposition for the reader who is not familiar with stochastic flows. Norris [A7] presents a very clear and complete account of Bismut's approach.

The proof of the existence of the density in the one-dimensional case can also be found in Zakai [A16].

The original approach of Malliavin [A5], [A6] has been developed by Stroock in [A11], [A12], [A13] and [A14]. Other expositions appear in Ikeda-Watanabe [A3], Kusuoka-Stroock [A4], Ocone [A8], Shigekawa [A9], [A10] and Watanabe [A15].

The result in 3.6.1 seems to be new. The regularity with respect to the starting point is studied in Stroock [A14] and the hypoellipticity of L in Stroock [A 14] and Bismut [A1].

3.4 AN APPLICATION TO NON LINEAR FILTERING: EXISTENCE AND REGULARITY OF THE CONDITIONAL DENSITY (following Bismut-Michel)

We consider in this paragraph, a "signal-observation" process (x, y) with values in $\mathrm{R}^n \times \mathrm{R}^p$, solution of a system:

$$dx_t = X_o(x_t)dt + X_i(x_t) \circ dw_t^i + Y_j(x_t) \circ dy_t^j$$
$$dy_t = h(x_t)dt + dv_t$$

where w and v are independant Wiener processes with values in R^m and R^p respectively, $X_0, X_1, \ldots, X_n,\ Y_1, \ldots, Y_p, h$ are in C_b^∞.

The problem is: does the filter associated to that system possess a smooth density, i.e. does there exist a smooth function $q(t,x)$ such that for every continuous function with compact support $\varphi : \mathrm{R}^n \to \mathrm{R}$:

$$\pi_t\varphi = E(\varphi(x_t)|\mathcal{Y}_t) = \int_{\mathrm{R}^n} \varphi(x)q(t,x)dx$$

where $\mathcal{Y}_t = \sigma(y_s, s \le t)$

If $h = 0$ and $Y_1 = \ldots = Y_p = 0$, the conditional law is the law of x and the answer to the question is in §3. We shall in a certain sense, go back to that situation by using Girsanov theorem and the theory of stochastic flows. But first remark that, heuristically, if $h = 0$ the conditional law is the law of the diffusion associated to the operator $\frac{1}{2}\Sigma X_i^2 + X_0 + \Sigma Y_i \frac{dy^i}{dt}$ whose drift term has $p+1$ independant components, namely $X_0, Y_1, \ldots, Y_p$. This is why it is natural to introduce the ideal $\mathcal{J}$ generated in $\mathcal{A} = LA(X_0, X_1, \ldots, X_m,\ Y_1, \ldots, Y_p)$ by the Lie algebra $\mathcal{B} = LA(X_1, \ldots, X_m)$.

3.4.1 The Girsanov theorem

$Z_t = \exp\left(\int_0^t h(x_S)dy_s - \frac{1}{2}\int_0^t |h(x_S)|^2 ds\right)$ is an exponential martingale and it allows to define a new probability measure $\overset{\circ}{P}$ on $(\Omega, \mathcal{F})$ by:

$$\frac{d\overset{\circ}{P}}{dP}|F_t = Z_t^{-1}, \quad t \ge 0.$$

then:

a. by Girsanov's theorem, y and w are independant Wiener processes under $\overset{\circ}{P}$.

b. by Kallianpur-Striebel's formula.

$$\pi_t\varphi \equiv E(\varphi(x_t)|\mathcal{Y}_t) = \frac{\overset{\circ}{E}(\varphi(x_t)Z_t|\mathcal{Y}_t)}{\overset{\circ}{E}(Z_t|\mathcal{Y}_t)} \equiv \frac{\sigma_t\varphi}{\sigma_t 1}$$

We get a first result in the uncorrelated case.

Theorem. Assume $Y_1 = \ldots = Y_p = 0$ and $\mathcal{J}(x) = \mathrm{R^n}$, $\forall \mathrm{x} \in \mathrm{R^n}$, then π_t has a density.

Proof: Define $\rho(t,x) = \overset{\circ}{E}(Z_t|x_t = x, Y_t)$, then, by the independance of x and y under $\overset{\circ}{P}$ we have:

$$\sigma_t\varphi = \overset{\circ}{E}(\varphi(x_t)\rho(t,x_t))$$

Under the assumption, the law of x_t, which is the same under P or $\overset{\circ}{P}$, has a density p and: $\sigma_t\varphi = \int \varphi(x)\rho(t,x)\,p(t,x)dx$. Therefore, π_t has the density:

$$q(t,x) = \frac{p(t,x)\rho(t,x)}{\int p(t,x)\rho(t,x)dx}.$$

3.4.2 The stochastic flows

In the case of correlated noises, more work is needed. In order to be able to fix the trajectory y in x_t, we write x_t as a composition of stochastic flows (cf §3.6.3).

Denote by ψ_t the stochastic flow associated to the following SDE.

$$du_t = Y_j(u_t) \circ dy_t^j$$

It is characterized by the following properties:

(i) $t \to \psi_t(y, u_0)$ is the essentially unique solution of the SDE

(ii) $(t, u_0) \to \psi_t(y, u_0)$ is a continuous map for every y, a.s.

(iii) $u_0 \to \psi_t(y, u_0)$ is a C^∞ diffeomorphism, $\forall (t, y)$, a.e.

Proposition. For $a.e. y, \bar{x}_t = \psi_t^{-1}(y, x_t)$ is the essentially unique solution of the SDE:

$$d\bar{x}_t = \psi_t^{*-1} X_i(\bar{x}_t) \circ dw_t^i + \psi_t^{*-1} X_0(\bar{x}_t)dt$$

where

$$\psi_t^{*-1}X(x) \equiv \left[\frac{\partial \psi_t}{\partial x}(x)\right]^{-1} X(\psi_t(x)).$$

Since ψ_t is a C^∞ diffeomorphism, to prove that the conditional law of x_t given y_t under $\overset{\circ}{P}$ has a smooth density, it suffices to prove that the law of $\bar{x}_t$ has a smooth density, for *a.e.y.* Of course, we cannot apply directly the results of §3 to the equation satisfied by $\bar{x}_t$ because it is inhomogeneous and with unbounded coefficients. But, by first assuming that the vector fields $Y_1, \ldots, Y_p$ have compact support which insures the boundedness of ψ_t and its derivatives and then passing to the limit, one can show that everything works in a similar way. If τ_t (resp. $\bar{\tau}_t$) is the stochastic flow associated to x_t (resp. $\bar{x}_t$), one gets:

$$D_s\bar{x}_t = \bar{\tau}_t^*\bar{\tau}_s^{*-1}(\psi_s^{*-1}X_i)(x_0)$$

Since $\tau_t = \psi_t \circ \bar{\tau}_t$, we have:

$$D_s\bar{x}_t = \bar{\tau}_t^*\tau_s^{*-1}X_i(x_0).$$

Since $\bar{\tau}_t^*$ is an invertible linear map, the invertibility of the Malliavin matrix is equivalent to that of:

$$C_t = \sum_{k=1}^{m}\int_0^t \left(\tau_s^{*-1}X_k(x_0)\right)\left(\tau_s^{*-1}X_k(x_0)\right)^t ds$$

Theorem. Assume $\mathcal{J}(x_0) = \mathrm{R}^n$. then C_t is a.s. invertible and π_t has a smooth density.

Proof: It is similar to that of Theorem 3.4.2. We just write the stochastic differential of $\tau_t^{*-1}X_i(x_0)$ which shows how the ideal $\mathcal{J}$ enters.

$$\begin{aligned} d\tau_t^{*-1}X_i(x_0) &= \tau_t^{*-1}[X_0, X_i](x_0)dt + \tau_t^{*-1}[X_j, X_i](x_0) \circ dw_t^j \\ &\quad + \tau_t^{*-1}[Y_j, X_i](x_0) \circ dy_t^j \end{aligned}$$

Remarks.

a. We have presented here a simplified version of the Bismut-Michel approach, which yields only the existence of a density for Π_t. Indeed, in order to obtain the smoothness of that density along the above lines, we would have to prove both the smoothness of the conditional density under $\overset{\circ}{P}$, and the smoothness o $x \to p(t,x)$. Bismut-Michel obtain the smoothness of the density of Π_t via a different and more direct approach.

b. The same kind of result can be obtained under more general assumptions.

- X_0, X_i, Y_j, h can depend on y_t.
- the coefficient of dv_t can be a function of $y_t : dy_t = h(x_t, y_t)dt + l(y_t) \circ dv_t$

3.4.3 Bibliographical comments

Our reference here has been Bismut-Michel [B1], which is an improvement over the original work of Michel [B7]. The same result can be obtained by an adaptation of the PDE hypoellipticity argument. For a summary of the two approaches and further references, we refer to Michel [B8]. Let us finally mention the works of Ferreyra [B4] and Kusuoka-Stroock [B6] who present other versions of the proof of existence and smoothness of the conditional law in nonlinear filtering, using the Malliavin calculus.

3.5 ANOTHER APPLICATION TO NON LINEAR FILTERING: MALLIAVIN CALCULUS APPLIED TO ZAKAI EQUATION AND NON EXISTENCE OF FINITE DIMENSIONAL FILTERS (following D. Ocone)

We consider a filtering problem as in §4 and assume that the signal noise and the observation noise are independent i.e. that the vector fields $Y_1 \ldots, Y_p$ are identically zero. So (x, y) is a solution of:

$$dx_t = X_0(x_t)dt + X_i(x_t) \circ dw_t^i$$

$$dy_t = h(x_t)dt + dv_t$$

and $X_0, X_1, \ldots, X_m, h$ are C^∞ and bounded as well as all their derivatives.

From §4, we know that, under Hörmander's condition $(\mathcal{J}(x_0) = \mathrm{R}^n)$, the unnormalized filter has a density p_t. Assume: $(X_1, \ldots, X_m)$ is a uniformly elliptic system of vector fields, i.e., there exists $d > 0$ such that

$$\forall v \in \mathrm{R}^m, \quad \sum_{i=1}^{m} (X_i, v)^2 \geq d\|v\|^2$$

then, Krylov Rozovskii and Pardoux observed that p_t is in $L^2(\Omega \times [0,T], H^{1,2}(\mathrm{R}^n)) \cap L^2(\Omega, C([0,T], L^2(\mathrm{R}^n)))$ and that it is the unique solution in this space of Zakai's equation

$$dp_t = L_0^* p_t dt + L_i^* p_t \circ dy_t^i$$

where

$$L_0 = \frac{1}{2}\left(\sum_{i=1}^{m} X_i^2\right) + X_0 - \frac{1}{2}h^2$$

$$L_i = h_i$$

From now on, we consider p_t as the unique process with values in $L^2(\mathrm{R}^n)$ solution of Zakai's equation.

The idea is to use Malliavin's calculus to determine whether this process "fills up" $L^2(\mathrm{R}^n)$ or not. Of course, we cannot speak of a density for the law of p_t in $L^2(\mathrm{R}^n)$ because there is no Lebesgue measure in $L^2(\mathrm{R}^n)$ but Ocone looks at all the finite dimensional projections of p_t in $L^2(\mathrm{R}^n)$ and finds a condition of Hörmarder's type under which every such projection has a density.
Denote by $\wedge$ the ideal of differential operators in R^n generated by Lie $(L_1^*, \ldots, L_p^*)$ in Lie $(L_0^*, L_1^*, \ldots, L_p^*)$ and, if $p \in L^2(\mathrm{R}^m)$, by $\wedge(p)$ the space $\{Dp, D \in \wedge\}$. Then, we have a result which is an infinite dimensional analog of that in §3.2.

3.5.1 Theorem.

Assume $\wedge p_0$ is dense in $L^2(\mathrm{R}^n)$. Then for any $t > 0$, the law of every finite dimensional projection of p_t has a density.

This result has an important consequence in terms of finite dimensional filters.

3.5.2 Definition.

For any $t > 0$, p_t is said to be finite dimensionally computable if there exists a finite dimensional random vector $Z(t, y)$ evolving in a finite dimensional vector space such that: $p_t(x; y) = F_t(x; Z(t, y))$ for some smooth function $F_t(\cdot)$.

3.5.3 Theorem.

If $\wedge p_0$ is dense in $L^2(\mathrm{R}^n)$, then for any t, p_t is not finite dimensionally computable.

Proof: Suppose p_t is finite dimensionally computable, then $F_t(\cdot\,; Z(t, y))$ stays in a finite dimensional sub manifold of $L^2(\mathrm{R}^n)$ when y evolves in $C([0, t], \mathrm{R}^p)$. So if S is a sub-space of $L^2(\mathrm{R}^n)$ of dimension larger than that of the state space of Z, then the projection of p_t on S cannot have a density, which contradicts Theorem 5.1.

3.5.4 Sketch of the proof of Theorem 3.5.1

Denote by S a finite dimensional subspace of $L^2(\mathrm{R}^n)$ and by π_S the orthogonal projection on S.

3.5.4.1 Lemma

$\forall t \in \mathrm{R}_+, \mathrm{p}_t$ is in the space $H^{2,1}$ of functionals from Ω into $L^2(\mathrm{R}^n)$ having a gradient.

This gradient is a Hilbert-Schmidt operator on $L^2(\mathrm{R}^n)$ whose kernel $D_s p_t$ satisfies;

$$D_s p_t = L_1^* p_s + \int_0^t L_0^* D_s p_r dr + \int_0^t L_1^* D_s p_r \circ dy_r$$

(we assume here for simplicity that the dimension of y is one).

Denote by $\Phi(t,s)$ the H.S. operator on $L^2(\mathrm{R}^n)$ solution of the linear equation: $\forall \varphi \in L^2(\mathrm{R}^n)$:

$$\Phi(t,s)\varphi = \varphi + \int_s^t L_0^* \Phi(r,s)\varphi dr + \int_s^t L_1^* \Phi(r,s)\varphi \circ dy_r$$

Then $D_s p_t = \Phi(t,s) L_1^* P_s$.

Let A be the Malliavin covariance matrix of p_t:

$$A_t = \int_0^t D_s p_t \otimes D_s p_t ds.$$

3.5.4.2 Lemma

If A_t is positive definite a.s., then the Malliavin covariance matrix associated to $\pi_S p_t$ is positive definite a.s., for every finite dimensional subspace S of $L^2(\mathrm{R}^n)$

Proof: Remark that the projection π_S and the gradient operator commute, ie.

$$D_S \pi_S p_t = \pi_S D_S p_t$$

So, if $A_{S,t}$ is the Malliavin covariance matrix associated to $\pi_S p_t$, one gets:

$$A_{S,t} = \pi_S A_t \pi_S^*.$$

and the result follows.

3.5.4.3 Lemma

If $\wedge p_0$ is dense in $L^2(\mathrm{R}^n)$, then A_t is positive definite a.s.

Proof: The difference with the finite dimensional case is that we cannot write $\Phi(r,s) = \Phi(r)\Phi(s)^{-1}$ because $\Phi(s)^{-1}$ would be an unbounded operator whose domain is not known. So, instead of working on $\Phi(s)^{-1}\varphi$ we work directly on $\langle A_t, \varphi, \varphi\rangle$ that we write as:

$$\int_0^t \sum_i < L_i^* p_s, \Phi^*(t,s)\varphi >^2 ds$$

Let $v(s,t,\varphi) = \Phi^*(t,s)\varphi$. Then, as a function of s, for (t,φ) fixed, v solves the following backward SPDE:

$$\begin{aligned} dv_s &= -L_0 v_s ds - L_i v_s \circ dy_s^i \\ v_t &= \varphi \end{aligned}$$

and the starting point of the proof is the following equality

$$\begin{aligned} \langle L_i^* p_s, \Phi^*(t,s)\varphi\rangle = \langle L_i^* p_0, \Phi^*(t,0)\varphi\rangle \\ + \int_0^s \langle [L_i^*, L_0^*] p_r, \Phi^*(t,r)\varphi\rangle dr \\ \int_0^s \langle [L_i^*, L_j^*] p_r, \Phi^*(t,r)\varphi\rangle \circ dy_r^j \end{aligned}$$

As in the classical case, this implies that, if $\langle A_t\varphi, \varphi\rangle = 0$, then $\Phi^*(t,0)\varphi$ is orthogonal to $\wedge p_0$. So: $\Phi^*(t,0)\varphi = v(0) = 0$. This implies $\varphi = 0$, from a backward uniqueness Theorem for parabolic PDEs.

Of course, what we need is stronger because we want the exceptional set in y to be independent of φ. This is done by using a pathwise interpretation and a technical Lemma.

Another difficulty we have overlooked is that v_t is not adapted: this can be solved by using either a pathwise interpretation or the extended stochastic calculus.

3.5.5 Bibliographical comments

The contents of this section is based on the work of Ocone [B 9]. Ocone has applied his results to the cubic sensor problem in [B 10].

Let us now explain in what sense the results of Ocone are stronger than earlier results about nonexistence of finite dimensional sufficient statistics, due in particular to Hazewinkel-Marcus [B5], Sussmann [B11] and Chaleyat Maurel-Michel [B3]. What the results of these authors preclude is the existence of a smooth finite dimensional manifold M, smooth vector fields over $M: Z_0, Z_1, \ldots, Z_p$, and a smooth mapping $F: \mathrm{R}_+ \times \mathrm{M} \to \mathrm{L}^2(\mathrm{R}^n)$ s.t.:

$$p_t = F(t, V_t), \qquad t \geq 0; \qquad \text{where}:$$
$$dV_t = Z_0(V_t)dt + \sum_{i=1}^{p} Z_i(V_t) \circ dy_t^i$$

On the other hand, Ocone shows that under appropriate conditions, for each fixed $t > 0$, there does not exist a finite dimensional manifold M, an M-valued random variable V_t, and a smooth mapping $F(t): M \to L^2(\mathrm{R}^n)$, such that for that t:

$$p_t = F(t, V_t).$$

We finally note that the stochastic calculus of variation has also been used by Chaleyat-Maurel [B2] in order to study the smoothness of the mapping $\{y(s); 0 \leq s \leq t\} \to p_t$.

3.6 EXTENDED STOCHASTIC INTEGRALS AND STOCHASTIC CALCULUS (Following Skorohod; Gaveau-Trauber; Nualart-Zakai; Nualart-Pardoux).

For simplicity, all processes in this section will be one dimensional, $(\Omega, \mathcal{F}, P)$ is the canonical space $C([0,1], \mathrm{R})$ equipped with Wiener measure and $w_t(\omega) = w(t)$.

3.6.1 The Skorohod integral

In this paragraph we develop the idea of Remark 3.2.3.

We saw in §2, that, if $\Phi \in H^{2,1}$ and $u \in L^2([0,1] \times \Omega, \mathrm{R})$ and is adapted, the following integration by parts formula holds:

$$E(\Phi \int_0^1 u_s dw_s) = E\left(\int_0^1 D_t\Phi \; u_t dt\right).$$

This can be rewritten as:

$$E(\Phi\delta u) = E((D\Phi, u))$$

and $u \longmapsto \delta u = \int_0^1 u_s dw_s$ appears as the adjoint of the gradient operator. Actually, we can extend the domain of δ to every process u such that the map:

$$\Phi \longrightarrow E((D\Phi, u))$$

is continuous on $L^2(\Omega)$.

It can be shown that this is the domain of definition of the Skorohod integral of u. Here, we shall take δ as a definition for the Skorohod integral.

3.6.1.1 Definition

(i) An element $u \in L^2((0,1) \times \Omega, \mathrm{R})$ is said to be Skorohod integrable if:

$$\exists C > 0, \forall \Phi \in H^{2,1}, |E((D\Phi, u))| \leq C\|\Phi\|_2$$

(ii) Let $\delta : L^2((0,1) \times \Omega, \mathrm{R}) \to \mathrm{L}^2(\Omega)$ be the unbounded operator with domain:

$$\text{Dom } \delta = \{\text{Skorohod integrable processes}\}$$

and defined by: if $u \in$ Dom δ, $\forall \Phi \in H^{2,1}, E(\Phi\, \delta\, u) = E((D\,\Phi, u))$

3.6.1.2 Basic properties of the Skorohod integral

Notation:

$$F_t = \sigma(w_s, s \leq t)$$

$$F^t = \sigma(w_s - w_t,\ t \leq s \leq 1)$$

Theorem.

a. If $u \in L^2((0,1) \times \Omega, \mathrm{R})$ and is F_t-adapted, then $u \in$ Dom δ and $\delta u = \int_0^1 u_s dw_s$ is the Itô integral.

b. If $u \in L^2((0,1) \times \Omega, H^{2,1})$, then $u \in$ Dom δ and $\delta u = \int_0^1 u_s dw_s$ is the backward Itô integral.

c. If $u \in L^{2,1} = L^2((0,1), H^{2,1})$, then $u \in$ Dom δ.

d. If $u \in$ Dom δ and $\Phi \in H^{2,1}$, then:

$$\delta(\Phi\, u) = \Phi\, \delta u - \int_0^1 D_t\Phi\ u_t dt$$

in the sense that $\Phi u \in \text{Dom}\, \delta$ if and only if the right hand side is square integrable and then the equality holds.

Proof: (a) is a consequence of Theorem 3.2.3, and (b) is proved in the same way. (d) follows from the definition of δ. Let us indicate a proof of (c), let $u \in L^{2,1}$. Define $t_k^n = k2^{-n}$,

$$u_{n,k} = \frac{1}{t_{k+1}^n - t_k^n} \int_{t_k^n}^{t_{k+1}^n} u_s ds$$

$$u_n(t) = \sum_{k=0}^{2^n-1} u_{n,k} 1_{[t_k^n, t_{k+1}^n L]}(t)$$

Consider the random variable:

$$\xi_n = \sum_{k=0}^{2^n-1} u_{n,k}(w_{t_{k+1}^n} - w_{t_k^n}) -$$
$$- \frac{1}{t_{k+1}^n - t_k^n} \int_{t_k^n}^{t_{k+1}^n} \int_{t_k^n}^{t_{k+1}^n} D_r u_s ds\, dr$$

It can be shown that $\xi_n \in L^2(\Omega)$; it then follows from (d) that $u_n \in \operatorname{Dom}\delta$ and $\delta u_n = \xi_n$. One can moreover show that $E(\xi_n \times \xi_m)$ converges as $n, m \to \infty$. Then $\{\xi_n, n \in rmN\}$ is Cauchy in $L^2(\Omega)$. But $u_n \to u$ in $L^2(\Omega \times (0,1))$, and δ is a closed operator (as the adjoint of the operator D which has a dense domain). Consequently, $u \in \operatorname{Dom}\delta$, and

$$\delta u = L^2 - \lim_{n\to+\infty} \delta u_n$$

3.6.1.3 Further properties

(i) In cases (a, b, c):

$$\delta u = L^2 - \lim_{n\to+\infty} \sum_{k=0}^{2^n-1} \bar{u}_{n,k}(w_{t_{k+1}^n} - w_{t_k^n})$$

where

$$\bar{u}_{n,k} = E\left(u_{n,k} | F_{t_k^n} \vee F^{t_{k+1}^n}\right)$$

(ii) In cases (a, b, c):

$$E(\delta u) = 0$$

and in case c:

$$E(\delta u)^2 = E\left(\int_0^1 u_t^2 dt\right) + E\left(\int_0^1 \int_0^1 D_s u_t D_t u_s ds\, dt\right)$$

Proof: (a) is a consequence of Theorem 3.2.3, and (b) is proved in the same way. (d) follows from the definition of δ. Let us indicate a proof of (c), let $u \in L^{2,1}$. Define $t_k^n = k2^{-n}$,

$$u_{n,k} = \frac{1}{t_{k+1}^n - t_k^n} \int_{t_k^n}^{t_{k+1}^n} u_s ds$$

$$u_n(t) = \sum_{k=0}^{2^n-1} u_{n,k} 1_{[t_k^n, t_{k+1}^n L]}(t)$$

Consider the random variable:

$$\xi_n = \sum_{k=0}^{2^n-1} u_{n,k}(w_{t_{k+1}^n} - w_{t_k^n}) -$$

$$- \frac{1}{t_{k+1}^n - t_k^n} \int_{t_k^n}^{t_{k+1}^n} \int_{t_k^n}^{t_{k+1}^n} D_r u_s ds\, dr$$

It can be shown that $\xi_n \in L^2(\Omega)$; it then follows from (d) that $u_n \in \operatorname{Dom}\delta$ and $\delta u_n = \xi_n$. One can moreover show that $E(\xi_n \times \xi_m)$ converges as $n, m \to \infty$. Then $\{\xi_n, n \in rmN\}$ is Cauchy in $L^2(\Omega)$. But $u_n \to u$ in $L^2(\Omega \times (0,1))$, and δ is a closed operator (as the adjoint of the operator D which has a dense domain). Consequently, $u \in \operatorname{Dom}\delta$, and

$$\delta u = L^2 - \lim_{n \to +\infty} \delta u_n$$

3.6.1.3 Further properties

(i) In cases (a, b, c):

$$\delta u = L^2 - \lim_{n \to +\infty} \sum_{k=0}^{2^n-1} \bar{u}_{n,k}(w_{t_{k+1}^n} - w_{t_k^n})$$

where

$$\bar{u}_{n,k} = E\left(u_{n,k} | F_{t_k^n} \vee F^{t_{k+1}^n}\right)$$

(ii) In cases (a, b, c):

$$E(\delta u) = 0$$

and in case c:

$$E(\delta u)^2 = E\left(\int_0^1 u_t^2 dt\right) + E\left(\int_0^1 \int_0^1 D_s u_t D_t u_s ds\, dt\right)$$

this last term vanishes in case a because if $s > t$, $D_s u_t = 0$, and in case b since if $s < t$, $D_s u_t = 0$.

3.6.2 Skorohod's integral as a process

Define: $\int_0^t u_s dw_s = \delta(u\, 1_{[0,t]})$ for $t \in (0,1)$.

3.6.2.1 Theorem.

The process $\{\int_0^t u_s dw_s,\ t \in [0,1]\}$ has a continuous modification if $u \in L^{2,1}$ and one of the following properties is satisfied:

- $\exists p > 1, \sup_{t\in[0,1]} E\left(\left(\int_0^1 |D_s u_t|^2 ds\right)^p\right) < +\infty$
- $\exists p > 2, E\left(\int_0^1 \left(\int_0^1 |D_s u_t|^2 ds\right)^p dt\right) < +\infty.$

3.6.2.2 Theorem.

Under the assumptions of Theorem 3.6.2.1,

$$\sum_{k=0}^{2^n-1} \left(\int_{t_k^n}^{t_{k+1}^n} u_s dw_s\right)^2 \longrightarrow \int_0^1 u_s^2\, ds$$

in probability, as $n \to \infty$.

3.6.3 The generalized Itô formula

Notation:

$H^{p,1} = \{\Phi \in L^p(\Omega), D\Phi \in L^p(\Omega \times (0,1))\}$

$L^{p,1} = L^p((0,1), H^{p,1})$

$L^{p,2} = \{u \in L^{p,1}, D_t u \in H^{p,1}$ a.e., $E\left(\int_0^1 \int_0^1 |D_s D_t u|^p ds\, dt\right) < +\infty\}$

Theorem. Consider:

$$X_t = X_0 + \int_0^t A_s\, ds + \int_0^t B_s\, dw_s$$

where: $X_0 \in H^{4,1}$, $A \in L^{4,1}$, $B \in L^{p,2}$ for some $p > 4$.

and let $F \in C^2(\mathrm{R})$. Then:

$$F(X_t) = F(X_0) + \int_0^t F'(X_s)\, A_s\, ds + \int_0^t F'(X_s) B_s\, dw_s$$
$$+ \frac{1}{2}\int_0^t F''(X_s)\, (\nabla X)_s\, B_s\, ds.$$

where:

$$(\nabla X)_t = \lim_{\substack{s \to t \\ s > t}} D_t X_s + \lim_{\substack{s \to t \\ s < t}} D_t X_s = D_t^+ X_t + D_t^- X_t$$
$$= B_t + 2\left(D_t X_0 + \int_0^t D_t X_r \, dr + \int_0^t D_t B_r \, dw_r\right)$$

Remark. In the adapted case, $(\nabla X)_t = B_t$ and this formula reduces to the usual Itô formula.

Proof: The three ingredients of the proof are the continuity of $\{X_t\}$ which follows from Theorem 6.2.1. Property 6.1.2. d, and Theorem 6.2.2. Write t_k for t_k^n.

$$F(X_t) - F(X_0) = \lim_n \sum_{k=0}^{[t2^n]-1} [F(X_{t_{k+1}}) - F(X_{t_k})]$$

$$= \lim_n \sum_k F'(X_{t_k}) \int_{t_k}^{t_{k+1}} A_s \, ds + \lim_n \sum_k F'(X_{t_k}) \int_{t_k}^{t_{k+1}} B_s \, dW_s +$$
$$+ \frac{1}{2} \lim_n \sum_k F''(X_k) \left(\int_{t_k}^{t_{k+1}} B_s \, dW_s\right)^2$$

clearly,

$$\sum_k F'(X_{t_k}) \int_{t_k}^{t_{k+1}} A_s \, ds \to \int_0^t F'(X_s) A_s \, ds$$

The convergence:

$$\frac{1}{2} \sum_k F''(X_k) \left(\int_{t_k}^{t_{k+1}} B_s \, dW_s\right)^2 \to \frac{1}{2} \int_0^t F''(X_s) B_s^2 ds$$

is an easy consequence of Theorem 6.2.2.

$$\sum_k F'(X_{t_k}) \int_{t_k}^{t_{k+1}} B_s dW_s = \sum_k \int_{t_k}^{t_{k+1}} F'(X_{t_k}) B_s dW_s +$$
$$+ \sum_k \int_{t_k}^{t_{k+1}} F''(X_{t_k}) D_s X_{t_k} B_s ds$$
$$\to \int_0^t F'(X_s) B_s dW_s + \frac{1}{2} \int_0^t F''(X_s)[(\nabla X)_s - B_s] B_s \, ds$$

3.6.4 The extended Stratonovitch integral

Definition. We say that $\{u_t, t \in (0,1)\}$ is Stratonovitch integrable if:

$$\sum_{k=0}^{2^n-1} u_{n,k} \left(w_{t_{k+1}^n \wedge t} - w_{t_k^n \wedge t}\right)$$

converges in probability, $\forall t \in [0,1]$. In that case the limit, denoted $\int_0^t u_s o dw_s$, is called the Stratonovitch integral of u.

Notation:

$L_c^{2,1} = \{u \in L^{2,1}/$ the set of functions $\{s \to D_t u_s;\ s \in [0,1] - \{t\}\}_{t\in[0,1]}$ is equi-continuous, and $\sup_{s,t} E(|D_t u_s|^2) < +\infty\}$.

If $u \in L_c^{2,1}$, one can define $D_t^+ u_t, D_t^- u_t$ and so $(\nabla u)_t$.

Theorem. If $u \in L_c^{2,1}$, u is Stratonovitch integrable and:

$$\int_0^t u_s \circ dw_s = \int_0^t u_s dw_s + \frac{1}{2}\int_0^t (\nabla u)_s\, ds.$$

Remark. If u_t is a F_t-semimartingale:

$$\int_0^1 u_t \circ dw_t = \int_0^1 u_t dw_t + \frac{1}{2}\langle u, w\rangle_1$$

If u_t is a F^t "backward semi-martingale"

$$\int_0^1 u_t \circ dw_t = \int_0^1 u_t dw_t - \frac{1}{2}\langle u, w\rangle_1.$$

More generally, if $u \in L_c^{2,1}$, $\langle u, w\rangle_1$ exists and is given by:

$$\langle u, w\rangle_1 = \int_0^1 (D_t^+ u_t - D_t^- u_t) dt$$

In the case u F_t adapted, $D_t^-\ u_t = 0$. So:

$$\begin{aligned}\frac{1}{2}\langle u, w\rangle_1 &= \frac{1}{2}\int_0^1 D_t^+\ u_t\ dt \\ &= \frac{1}{2}\int_0^1 (\nabla u)_t\ dt\end{aligned}$$

In the case u F^t adapted,

$D_t^+ \mu_t = 0$. So:

$$\begin{aligned}\frac{1}{2}\langle u, w\rangle_1 &= \frac{1}{2}\int_0^1 D_t^- u_t\, dt \\ &= \frac{1}{2}\int_0^t (\nabla u)_t\, dt\end{aligned}$$

3.6.5 The generalized Stratonovich formula

Let the assumptions of Itô's formula be in force and assume in addition:

$$B \in L_c^{2,1} \qquad \text{and} \qquad \nabla B \in L^{4,1}$$

then, if: $X_t = X_0 + \int_0^t A_s ds + \int_0^t B_s \circ dw_s$,

$$F(X_t) = F(X_0) + \int_0^t F'(X_s)A_s ds + \int_0^t F'(X_s)B_s \circ dw_s$$

3.6.6 Bibliographical comments

We have mainly followed the presentation and results of Nualart-Pardoux [C 4]. Earlier results on the Skorohod integral include the work of Skorohod [C 14], Gaveau-Trauber [C 1], Krée [C 2] and Nualart-Zakai [C 5]. Note that our notion of "Skorohod integrable processes" is slightly more general than that in the original paper by Skorohod. Various versions of the generalized Itô formula have been given by Sevljakov [C 12], Sekiguchi-Shiota [C 11], Ustunel [C 15] and Nualart-Pardoux [C 4]. Ogawa [C 8] has constructed a generalized stochastic integral which essentially coincides with what we call the "extended Stratonovich integral." For an overwiew and comparison of all the existing results, we refer the reader to the expository paper of Nualart [C 3]. For applications to stochastic differential equations, we refer to Ogawa [C 9], Shiota [C 13], Ocone-Pardoux [C 6], [C 7] and Pardoux-Protter [C 10].

REFERENCES

A. *Stochastic calculus of variations and its application to solutions of stochastic differential equations.*

[A1] BISMUT, J. M., (1981). ***Martingales, The Malliavin calculus and hypoellipticity under general Hörmander's conditions.*** Z. Wahrs, V. Geb. 56, 469-505.

[A2] BISMUT, J. M., (1982). ***An introduction to the stochastic calculus of variations,*** in Stochastic Differential Systems, M. Kohlmann & N. Christopeit eds, p.33-72, Lecture Notes in Control and Inf. Sci. 43, Springer-Verlag.

[A3] IKEDA, N. and WATANABE, S., (1981). ***Stochastic differential equations and diffusion processes.*** North Holland, Kodansha, Amsterdam/Tokyo.

[A4] KUSUOKA, S., and STROOCK, D. W., (1982). ***Applications of the Malliavin calculus.*** Part I. Taniguchi Symposium, Takata. K. Itô ed. Kinokuniya-Tokyo (1984) 277-306. Part II, J. Fac. Sci. Tokyo, Sec IA, 32 (1985) 1-76.

[A5] MALLIAVIN, P., (1976). ***Stochastic calculus of variation and hypoelliptic operators*** in Proc. Int. Symp. on SDE Kyoto K. Itô ed. p. 195-263, Kinokuniya-Tokyo (1978).

[A6] MALLIAVIN, P, (1978). ***C^k-hypoellipticity with degeneracy*** in Stochastic Analysi, A. Friedman and M. Pinsky eds. p. 199-214 and p. 327-340. Academic Press.

[A7] NORRIS, J, (1986). ***Simplified Malliavin Calculus,*** in Séminaire de Probabilités XX, J. Azéma, M. Yor Eds, Lecture Notes in Mathematics 1204, Springer Verlag, 101-130.

[A8] OCONE, D. (1988). ***A guide to the stochastic calculus of variations,*** Proc. Silivri Conf, Lecture Notes in Math. Springer-Verlag .

[A9] SHIGEKAWA, I., (1978). ***Absolute continuity of probability laws of Wiener functionals.*** Proc. Jap. Acad. 54, A, p.230-233.

[A10] SHIGEKAWA, I., (1980). ***Derivatives of Wiener functionals and absolute continuity of induced measures.*** J. Math. Kyoto Univ. 20. 263-289.

[A11] STROOCK, D. W., (1981). *The Malliavin calculus and its applications to second order parabolic differential operators.* Parts I and II. Math. Systems Theory, 14, 25-65 and 141-171.

[A12] STROOCK, D. W., (1981). *The Malliavin calculus. A functional analytic approach.* J. Funct. Analysis 44, 212-257.

[A13] STROOCK, D. W., (1983). *Some applications of stochastic calculus to partial differential equations.* Ecole d'été de Probabilités de St Flour XI. Lecture Notes in Mathematics 976, Springer-Verlag.

[A14] STROOCK, D. W., (1986). *Malliavin calculus and applications to PDEs.* IMA Preprint 278, IMA Minneapolis, MN.

[A15] STROOCK, D. W. and VARADHAN, S.R.S., *On the support of diffusion processes with applications to the maximum principle.* Proceedings of sixth Berkeley Symposium on Mathematical Statistics and Probability, Vol. III, 333-360.

[A16] WATANABE, S. *Lectures on stochastic differential equations and Malliavin calculus.* Tata Institute of Fundamental Research, Springer Verlag (1984).

[A17] ZAKAI, M. (1985). *The Malliavin calculus.* Acta Applicandae Mat. 3, 175-207.

B. Applications of the stochastic calculus of variations to nonlinear filtering.

[B1] BISMUT, J. M. and MICHEL, D., (1981). *Diffusions conditionnelles.* J. Funct. Anal. Part I: 44, 174-211, Part II: 45 (1982) 274-282.

[B2] CHALEYAT-MAUREL, M., (1986). *Robustesse en théorie du filtrage non linéaire et calcul des variations stochastique.* J. Funct-Anal.

[B3] CHALEYAT-MAUREL, M., - MICHEL, D., (1984) *Des résultats de non existence de filtre de dimension finie.* Stochastics 13.

[B4] FERREYRA, G., *Smoothness of the unnormalized conditional measure in stochastic nonlinear filtering.* Preprint Math. Dep. LSU Baton Rouge, LA, USA.

[B5] HAZEWINKEL, M. and MARCUS, S.,(1982). *On Lie Algebras and finite dimensional filtering.* Stochastics 7.

[B6] KUSUOKA, S. and STROOCK, D. W., (1984). *The partial Malliavin calculus and its application to nonlinear filtering.* Stochastics, 12, 83-142.

[B7] MICHEL, D., (1981). *Régularité des lois conditionelles en théorie du filtrage non linéaire et calcul des variations stochastique.* J. Funct. Anal. 41, 8-36.

[B8] MICHEL, D., (1984). *Conditional laws and Hörmander condition.* in Stochastic Analysis, K. Itô ed., North Holland, 387-408.

[B9] OCONE, D., *Stochastic calculus of variations for stochastic partial differential equations.* J. Funct, Anal., to appear.

[B10] OCONE, D., (1988). *Probability densities for conditional statistics in the cubic sensor problem.* Math. of Systems, Signals and Control.

[B11] SUSSMANN, H.,(1981). *Rigorous results on the cubic sensor problem.* in Stochastic Systems, Hazewinkel M. and Willems S. eds, Reidel.

C. Skorohod's integral and non adapted stochastic calculus

[C1] GAVEAU, B. and TRAUBER, P., (1982) *"L'intégrale Stochastique comme opérateur de divergence dans l'espace fonctionnel.* J. Funct. Anal. 46, 230-238.

[C2] KREE, M.,(1977). *Propriétés de trace en dimension infinie d'espaces de type Sobolev.* Bull. Soc. Math. France 105, 141-163.

[C3] NUALART, D., *Non-causal stochastic integrals and calculus.* Proc. Silivri Conf, Lecture Notes in Mathematics, Springer-Verlag (to appear).

[C4] NUALART, D. and PARDOUX, E., *Stochastic calculus with anticipating integrands.* Prob. Theory and Rel. Fields, to appear.

[C5] NUALART, D. and ZAKAÏ, M., (1986). *Generalized stochastic integrals and the Malliavin calculus.* Prob. Theory and Rel. Fields 73, 255-280.

[C6] OCONE, D. and PARDOUX, E. *A generalized Itô-Ventzell formula. Application to a class of anticipating stochastic differential equations.* Preprint.

[C7] OCONE, D. and PARDOUX, E. *Bilinear stochastic differential equations with boundary conditions.* Preprint.

[C8] OGAWA S., (1984). *Quelques propriétés de l'intégrale stochastique de type non-causal.* Japan J. of Applied Math. 1, 405-416.

[C9] OGAWA, S., (1984). *Sur la question d'existence de solutions d'une équation différentielle stochastique de type non-causal.* J. Math. Kyoto Univ. 24, 699-704.

[C10] PARDOUX, E. and PROTTER, P., *Stochastic Volterra equation with anticipating coefficients.* Preprint.

[C11] SEKIGUCHI, T. and SHIOTA, Y.(1985). *L^2 theory of non-causal stochastic integrals.* Math. Rep. Toyama Univ. 8, 119-195.

[C12] SEVLJAKOV, A., (1981). *The Itô formula for the extended stochastic integral.* Theory Prob. and Math. Stat. 22, 163-174.

[C13] SHIOTA, Y., (1986). *A linear stochastic integral equation containing the extended Itô integral.* Math. Rep. Toyama Univ. 9, 43-65.

[C14] SKOROHOD, A. V., (1975). *On a generalization of a stochastic integral.* Theory Prob. and Applic. 20, 219-233.

[C15] USTUNEL, A. S., (1986). *La formule de changement de variable pour l'intégrale anticipante de Skorohod.* CRAS.

4

STOCHASTIC CALCULUS ON MANIFOLDS WITH APPLICATIONS

Tyrone Duncan

INTRODUCTION

This material is based on some lectures on stochastic calculus in manifolds that were given by the author at the Systems Research Center of the University of Maryland, College Park. Since stochastic calculus in manifolds has been a useful tool in a number of diverse areas, a broad spectrum of applications are presented in addition to the foundations of stochastic calculus in manifolds. Initially some results from differential geometry are described that are particularly important for stochastic calculus. The constructions of Brownian motion and its horizontal lift to the bundle of orthonormal frames for a compact Riemannian manifold are described. The extension of these constructions to noncompact manifolds is discussed. Some notions of real-valued stochastic integrals for a manifold-valued Brownian motion are given. Since an important emphasis of this material is the applications of stochastic calculus in manifolds, a wide range of applications are described. Initially three applications to differential geometry are given. These include a local formula for an index theorem, a result on the nonexistence of harmonic 1-forms and the description of a theta function for a compact Lie algebra as an integration of Wiener measure in a compact Lie group. Subsequently two applications to probability are given. These include an explicit expression for a Radon-Nikodym derivative for a measure absolutely continuous with respect to Wiener measure in a manifold and a law of the iterated logarithm for a manifold-valued Brownian motion. Some applications to stochastic control and filtering are given. These include the existence of an optimal control for a controlled diffusion process in a manifold satisfying a convexity condition and a brief description of a necessary and sufficient condition for an optimal control of a partially observed controlled diffusion process in a manifold. A problem of stochastic filtering in a manifold is discussed where the observation process is in a manifold and the signal process that appears in the observation process takes values in a Euclidean space. A stochastic equation is given for the evolution of the conditional expectation of a function of the signal process. The control and the stochastic filtering of discontinuous processes in a vector bundle where the discontinuities of the process occur in the fibres of the vector bundle are also briefly discussed. Two families of explicitly solvable stochastic control problems in manifolds are given. The first family are stochastic systems with values in the real hyperbolic spaces which are noncompact manifolds with constant negative sectional curvature. The second family are stochastic systems with values in the spheres. Finally an estimation problem in real hyperbolic three space is solved. This problem is to find explicitly the conditional probability density of a random initial condition of the Brownian motion in real hyperbolic three space given the observation of the Brownian motion at some fixed positive time. This material is concluded with some notes to provide the history and a guide for additional reading.

4.1 SOME RESULTS FROM DIFFERENTIAL GEOMETRY

Initially some results from differential geometry are briefly reviewed to isolate the important concepts for stochastic calculus and to fix the notation.

A pseudogroup of transformations on a topological space S is a set Γ of transformations satisfying the following axioms:

i) Each $f \in \Gamma$ is a homeomorphism of an open set of S onto another open set of S

ii) If $f \in \Gamma$ then the restriction of f to an arbitrary open subset of the domain of f is in Γ

iii) Let $U = \cup_i U_i$ where each U_i is an open set of S. A homeomorphism f of U onto an open set of S belongs to Γ if the restriction of f to U_i is in Γ for every i

iv) For each open set U of S, the identity transformation of U is in Γ

v) If $f \in \Gamma$ then $f^{-1} \in \Gamma$

vi) If $f_i \in \Gamma$ is a homeomorphism of U_i onto V_i for $i = 1, 2$ and if $V_i \cap U_2 \neq \phi$ then the homeomorphism $f_2 \circ f_1$ of $f_1^{-1}(V_1 \cap U_2)$ onto $f_2(V_1 \cap U_2)$ is in Γ.

An atlas of a topological space M compatible with a pseudogroup Γ is a family of pairs (U_2, φ_i) called charts such that

i) Each U_i is an open set of M and $\cup_i U_i = M$.

ii) Each φ_i is a homeomorphism of U_i onto an open set of S.

iii) If $U_i \cap U_j \neq \phi$ then the mapping $\varphi_j \circ \varphi_i^{-1}$ of $\varphi_i(U_i \cap U_j)$ onto $\varphi_j(U_i \cap U_j)$ is an element of Γ.

A complete atlas of M compatible with Γ is an atlas of M compatible with Γ that is not contained in any other atlas of M compatible with Γ. It is clear that complete atlases exist.

A differentiable manifold of class C^r is a Hausdorff space with a fixed complete atlas compatible with $\Gamma^r(\mathbb{R}^n)$ which is the pseudogroup of transformations of class C^v of $\mathbb{R}^n$. In this paper it is assumed that the manifolds are of class C^∞ unless otherwise stated. For our applications it usually suffices though to assume that the manifolds are of class C^2.

In the stochastic analysis that is described here the notion of a principal (frame) bundle plays a central role. A (differentiable) principal fibre bundle over a manifold M with Lie group G is a manifold P and an action of G on P satisfying the following:

i) G acts freely in P on the right: $(u, a) \in P \times G \longrightarrow ua = R_a u \in P$ where R_a is right translation by a.

ii) $M = P/G$ and the canonical projection $\pi : P \to M$ is differentiable.

iii) P is locally trivial.

P is the total space or the bundle space, M is the base space, G is the structure group and π is the projection. $\pi^{-1}(x)$ is the fibre over $x \in M$ and each fibre is diffeomorphic to G.

Principal bundles play a central role in the study of manifolds because other bundles can be constructed from them as associated bundles. Let $P(M, G)$ be a principal fibre bundle and let F be a manifold on which G acts on the left. Consider the group action

$$(u, \xi) \in P \times F \overset{a \in G}{\longmapsto} (u\, a, a^{-1}\, \xi) \in P \times F \tag{4.1}$$

$E(M, F, G, P) = P \times_G F$ is a fibre bundle associated with P with standard fibre F that is the quotient space $P \times F$ by the group action (1).

Principal bundles in this paper usually appear as frame bundles. A linear frame u at $x \in M$ is an ordered basis $X_1, \ldots, X_n$ of the n dimensional tangent space $T_x M$. The bundle of linear frames of $M, L(M)$, is the family of all linear frames at all points of M. Let $\pi : L(M) \rightarrow M$ be the canonical projection. The general linear group acts on $L(M)$ on the right. If $a = (a_i^j) \in GL(n, \mathbb{R})$ and $u = (X_1, \ldots, X_n)$ is a frame at $x \in M$ then $ua = (Y_1, \ldots, Y_n)$ where $Y_i = \sum_j a_i^j X_j$. If $(x^1, \ldots, x^n)$ is a local coordinate system for $U \subset M$ then $X_i = \sum X_i^k \left(\frac{\partial}{\partial x^k}\right)$ gives the frame $(X_1, \ldots, X_n)$ so that (x^j) and (X_i^k) are local coordinates in $\pi^{-1}(U)$ where $(X_i^k) \in GL(n, \mathbb{R})$.

The tangent bundle TM over M is the bundle associated with $L(M)$ with standard fibre $\mathbb{R}^n$.

The structure group G of a principal fibre bundle $P(M, G)$ is reducible to a Lie subgroup G' if and only if there is an open covering (U_α) of M with a set of transition functions that take their values in G'. For a manifold M with a Riemannian metric the bundle of linear frames can be reduced to the bundle of orthonormal frames.

For problems of stochastic analysis it is necessary to impose additional structure on a smooth manifold. One of these properties is a rule for differentiation. This is called a connection because it allows one to connect different fibres of the tangent bundle.

A connection Γ in a principal bundle $P = P(M, G)$ is an assignment of a subspace Q_u of $T_u P$ to each $u \in P$ such that

i) $T_u P = G_u \oplus Q_u$ where G_u is the subspace of $T_u P$ with vectors tangent to the fibre through u

ii) $Q_{ua} = (R_a)_* Q_u$ for each $u \in P$ and $a \in G$ where R_a is the transformation of P induced by $a \in G$ as $R_a u = ua$

iii) Q_u depends differentiably on u.

The second property above means that the distribution $u \longmapsto Q_u$ is G invariant. G_u is called the vertical subspace and Q_u is the horizontal subspace.

Each $A \in g$, the Lie algebra of G, induces a vector field A^* on the principal bundle $P(M, G)$ that is tangent to the fibres because the action of G on P induces a homomorphism σ of the Lie algebra g of G into the Lie algebra of vector fields on P. A^* is called the fundamental vector field corresponding to A. The map $A \longmapsto (A^*)_u$

is a linear isomorphism of g onto G_u for each $u \in P$. We define a 1-form ω on P with values in g by the rule that $\omega(X)$ is the unique $A \in g$ such that $(A^*)_u$ is the vertical component of X. ω is called the connection form of the connection Γ. It is elementary to verify that

$$\omega(A^*) = A \qquad \text{for each} \quad A \in g \tag{4.2}$$

and

$$(R_a)^*\omega = \mathrm{Ad}\,(a^{-1})\omega \tag{4.3}$$

for each $a \in G$ where $Ad(\cdot)$ is the adjoint representation of G in g. Given a g-valued 1-form that satisfies (2-3) there is a unique connection Γ in P whose connection form is ω.

The horizontal lift of a vector field X on M is a vector field X^* on P which is horizontal and which projects onto X. A horizontal curve in P is a piecewise differentiable curve of class C^1 whose tangent vectors are all horizontal. Let $\tau = x_t, t \in [a,b]$ be a piecewise differentiable curve of class C^1 in M. A horizontal lift of τ is a horizontal curve $\tau^* = u_t, t \in [a,b]$ in P such that $\pi(u_t) = x_t$, $t \in [a,b]$. Given a smooth curve in M it is possible to construct a horizontal lift.

Proposition 1. Let $\tau = x_t, t \in [0,1]$ be a curve of class C^1 in M. For an arbitrary point u_0 of P with $\pi(u_0) = x_0$ there is a unique horizontal lift $\tau^* = u_t, t \in [0,1]$ of τ which starts from u_0.

To verify this result it suffices initially to choose some lift $v_t, t \in [0,1]$ with $v_0 = u_0$. Any lift u_t of τ is related to v_t by the equation

$$u_t = v_t\, a_t \tag{4.4}$$

where a_t is a curve in G. Since

$$\dot{u}_t = \dot{v}_t\, a_t + v_t\, \dot{a}_t \tag{4.5}$$

it follows that

$$\omega(\dot{u}_t) = \mathrm{Ad}\,(a_t^{-1})\omega(\dot{v}_t) + a_t^{-1}\dot{a}_t \tag{4.6}$$

If u_t is a horizontal lift so that $\omega = \dot{u}_t$ it is necessary and sufficient that

$$\dot{a}_t\, a_t = -\omega(\dot{v}_t) \tag{4.7}$$

Thus solving this equation with $a_0 = e$ gives the unique horizontal lift.

The notion of curvature is easily defined from the connection form.

$$D\omega = (d\omega)h \tag{4.8}$$

is the g-valued two form that is called the curvature form of ω where d is exterior derivative and $h : T_uP \to Q_u$ is the projection. A connection in P is flat if and only if the curvature form vanishes dentically.

For a connection in a vector bundle an equivalent notion is covariant derivative. Let $P(M, G)$ be a principal fibre bundle and ρ a representation of G in $GL(m; F)$ where F is $\mathbb{R}$ or $\mathbb{C}$. Let $E(M, F^m, G, P)$ be the associated bundle with standard fibre F^m on which G acts through ρ. E is a vector bundle. Let φ be a section of E defined over the M-valued curve x_t. For each fixed t the covariant derivature $\nabla_{\dot{x}_t}\varphi$ of φ in the direction (or with respect to) $\dot{x}_t$ is defined by

$$\nabla_{\dot{x}_t}\varphi = \lim_{h\to 0}\frac{1}{h}\left[\tau_t^{t+h}(\varphi(x_t + h)) - \varphi(x_t)\right] \tag{4.9}$$

where $\tau_t^{t+h} : \pi_E^{-1}(x_{t+h}) \to \pi_E^{-1}(x_t)$ is the parallel displacement along $\tau = x_s$, $s \in [t, t+h]$ from x_{t+h} to x_t.

The parallel displacement of fibres of a principal bundle $P(M, G)$ can be defined from the horizontal lift of a smooth curve in M. Let $\tau = x_t, t \in [0,1]$ be a differentiable curve of class C^1 in M. Let $u_0 \in P$ be arbitrary such that $\pi(u_0) = x_0$. The unique horizontal lift τ^* of τ through u_0 has the end point u_1 such that $\pi(u_1) = x_1$. By varying $u_0 \in \pi^{-1}(x_0)$ we obtain a mapping of $\pi^{-1}(x_0)$ onto $\pi^{-1}(x_1)$ which maps u_0 into u_1. This mapping is called the parallel displacement along the curve τ. It is clear that parallel displacement along τ commutes with the action of G on P.

If $X \in T_xM$ and φ is a cross section of E defined in a neighborhood of x then the covariant derivature $\nabla_X\varphi$ of φ in the direction X is

$$\nabla_X\varphi = \nabla_{\dot{x}_0}\varphi \tag{4.10}$$

where $\tau = x_t, t \in (-\epsilon, \epsilon)$ is a curve such that $X = \dot{x}_0$.

Let $X, Y \in T_xM$ and let φ and ψ be cross sections of E defined in a neighborhood of x. The following properties of the covariant derivative are easy to verify:

i) $\nabla_{X+Y}\varphi = \nabla_X\varphi + \nabla_X\varphi$

ii) $\nabla_X(\varphi + \psi) = \nabla_X\varphi + \nabla_X\psi$

iii) $\nabla_{\lambda X}\varphi = \lambda \cdot \nabla_X\varphi$ where $\lambda \in \mathbb{R}$

iv) $\nabla_X(\lambda\varphi) = \lambda(x) \cdot \nabla_x\varphi + (X\lambda) \cdot \varphi(x)$ where λ is a real-valued function defined in a neighborhood of x.

A linear frame $u = (X_1, \ldots, X_n)$ at $x \in M$ can be given as $u : \mathbb{R}^n \to T_xM$ such that $ue_i = X_i$ for $i = 1, 2, \ldots, n$ where $(e_i, \ldots, e_n)$ is the canonical basis in $\mathbb{R}^n$. Thus a linear frame u at $x \in M$ can be defined as a nonsingular linear mapping of $\mathbb{R}^n$ onto T_xM. The canonical form Θ of P is the $\mathbb{R}^n$-valued 1-form defined by

$$\Theta(X) = u^{-1}(\pi(X)) \tag{4.11}$$

where $X \in T_uP$ and $u \in \mathrm{Hom}\,(\mathbb{R}^n, T_{\pi(u)}M)$ as defined above.

A connection in $P = L(M)$ is called a linear connection of M. If $\xi \in \mathbb{R}^n$ then there is a horizontal vector field $B(\xi)$ on P such that $\pi((B\,\xi)_u) = u(\xi)$. $B(\xi)$ is unique and is called the standard horizontal vector field corresponding to ξ.

Let $\mathbf{A}^n$ be affine n space. The group $A(n; \mathbb{R})$ of affine transformations of $\mathbf{A}^n$ is the family of matrices of the form

$$\tilde{a} = \begin{pmatrix} a & \xi \\ 0 & 1 \end{pmatrix} \tag{4.12}$$

where $a = (a^i_j) \in GL(n; \mathbb{R})$ and $\xi \in \mathbb{R}^n$. An affine frame at $x \in M$ consists of a point $p \in A_xM$ and a linear frame $(X_1, \ldots, X_n)$ at x. An affine frame $(p; X_1, \ldots, X_n)$ can be identified with an affine transformation $\tilde{u} : \mathbf{A}^n \to A_xM$.

Let $\tau = x_t, t \in [0,1]$ in M be a piecewise C^1 curve and let $Y_t = \tau_0^t(\dot{x}_t)$, $t \in [0,1]$ where τ_0^t is the linear parallel displacement along τ from x_t to x_0. Let $C_t, t \in [0,1]$ be the curve in the affine tangent space $A_{x_0}M$ that starts from the origin. C_t is called the development of τ into $A_{x_0}M$. The development of a curve $\tau = x_t, t \in [0,1]$ into $A_{x_0}M$ is a line segment if and only if the vector fields $\dot{x}_t$ along $\tau = x_t$ are parallel.

A curve $\tau = x_t, t \in (a,b)$ of class C^1 in M with a linear connection is a geodesic if the vector field $X = \dot{x}_t$ defined along τ is parallel along τ, that is, $\nabla_X X = 0$ for all $t \in (a,b)$. A curve is a geodesic if and only if it is the projection onto M of an integral curve of a standard horizontal vector field of $L(M)$. Given $x \in M$ and $X \in T_xM$ there is a unique geodesic with initial condition (x, X).

A linear connection is complete if every geodesic can be extended to a geodesic defined for $t \in \mathbb{R}$.

For some applications of differential geometry it is useful to have local coordinate descriptions of the previous notions.

Let M be a manifold and U a coordinate neighborhood in M with a local coordinate system $x^1, \dots, x^n$. The vector fields $\frac{\partial}{\partial x^i}$ for $i = 1, \dots, n$ defined in U are denoted X_i. Since each linear frame at a point $x \in U$ can be uniquely expressed by

$$\left(\sum_i X_1^i (X_i)_x, \dots, \sum_i X_n^i (X_i)_x\right) \tag{4.13}$$

where $\det(X_i^j) \neq 0$ so that (x^i, X_k^j) is a local coordinate system in $\pi^{-1}(U) \subset L(M)$. Let $Y = (Y_j^i)$ be the inverse matrix of (X_j^i).

Let $e_1, \dots, e_n$ be the canonical basis for $\mathbb{R}^n$ and set

$$\Theta = \sum_i \Theta^i e_i \tag{4.14}$$

Θ is the canonical 1-form which can be expressed as

$$\Theta^i = \sum_j Y_j^i dx^j \tag{4.15}$$

where (x^i, X_k^j) is the local coordinate system of $\pi^{-1}(U)$.

The connection form ω can be expressed as

$$\omega = \sum_{i,j} \omega_j^i E_i^j \tag{4.16}$$

where (E_i^j) is a basis of $gl(n; \mathbb{R})$, the Lie algebra of $GL(n; \mathbb{R})$. If σ is the cross section of $L(M)$ over U which assigns to $x \in U$ the linear frame $((X_1)_x, \dots, (X_n))_x$ then define ω_u as

$$\omega_U = \sigma^* \omega \tag{4.17}$$

Thus ω_U is a $gl(n; \mathbb{R})$-valued 1-form on U. Define the functions Γ_{jk}^i for $i, j, k = 1, 2, \dots, n$ on U by

$$\omega_U = \sum_{i,j,k} (\Gamma_{jk}^i dx^j) E_i^k \tag{4.18}$$

The functions Γ^i_{jk} are called the Christoffel symbols of the linear connection Γ with respect to the local coordinate system $x^1, \ldots, x^n$. The Christoffel symbols appear in the covariant derivative as

$$\nabla_{X_i} X_j = \sum_k \Gamma^k_{ij} X_k \tag{4.19}$$

A curve $\tau = x_t$ of class C^2 is a geodesic if and only if

$$\frac{d^2x^i}{dt^2} + \sum_{j,k} \Gamma^i_{jk} \frac{dx^j}{dt} \frac{dx^k}{dt} = 0 \tag{4.20}$$

for $i = 1, 2, \ldots, n$.

Let M be a manifold with a linear connection. For $X \in T_xM$ let $\tau = x_t$ be the geodesic with the initial condition (x, X). Define

$$\operatorname{Exp}(tX) = x_t \tag{4.21}$$

Identifying $x \in M$ with the zero vector in T_xM we can consider M as a submanifold of TM. Then Exp is a local diffeomorphism of a neighborhood of the zero vector in T_xM to a neighborhood of $x \in M$.

Let $u = (X_1, \ldots, X_n)$ be a linear frame at x. The linear isomorphism

$$u : \mathbb{R}^n \to T_xM$$

defines a coordinate system in T_xM. The composition of u with Exp defines a local coordinate system for a neighborhood of $x \in M$. This is called the normal coordinate system determined by the frame u. For this normal coordinate system $x^1, \ldots, x^n$ a geodesic with initial condition (x, X) where $X = \sum_i a_i X_i$ is given by

$$x^i = a^i t \qquad i = 1, \ldots, n \tag{4.22}$$

A Riemannian metric is a positive definite, symmetric, covariant tensor field of degree 2. If g is a Riemannian metric on M then the arc length of a differentiable curve $\tau = x_t$, $t \in [a, b]$ of class C^1 is

$$L = \int_a^b \left(\sum g_{ij} \frac{dx^i}{dt} \frac{dx^j}{dt} \right)^{\frac{1}{2}} dt \tag{4.23}$$

This definition is generalized to a piecewise differentiable curve of class C^1 in an obvious way. The distance $d(x, y)$ between $x, y \in M$ is defined as the infimum of the lengths of all piecewise differentiable curves of class C^1 joining x and y. Given a Riemannian metric on M the Riemannian connection is the unique connection such that g is parallel with respect to the connection and the torsion is zero. For our purposes a Riemannian manifold is a manifold with a Riemannian metric and the Riemannian connection. For a Riemannian manifold the structure group $GL(n; \mathbb{R})$ of $L(M)$ can be reduced to the orthogonal group $O(n)$ and one has the reduced principal bundle $O(M)$, the bundle of orthonormal frames.

Let U be a normal coordinate neighborhood of x with a normal coordinate system $X^1, \ldots, x^n$ at x. A cross section σ of $O(M)$ over U is defined. Let u be the orthonormal frame at x given by $(\frac{\partial}{\partial x^1})_x, \ldots, (\frac{\partial}{\partial x^n})_x$. By the parallel displacement of u along the geodesics through x, an orthonormal frame is attached to every point of U. For Riemannian manifolds this cross section of $O(M)$ is often more useful then the cross section of $L(M)$ induced by the local coordinates $x^1, \ldots, x^n$.

For $r > 0$ sufficiently small let $U(x; r)$ be the neighborhood of $x \in M$ defined by $\sum_i (x^i)^2 < r^2$. There is an $a > 0$ such that if $\rho \in (0, a)$ then any two points of $U(x; \rho)$ can be joined by a unique minimizing geodesic which lies in $U(x; \rho)$. $U(x; \rho)$ is called a convex neighborhood of x.

A Riemannian manifold is said to be complete if the Riemannian connection is complete, that is, if every geodesic of M, Exp tX, is defined for all $t \in \mathbb{R}^1$. For a connected Riemannian manifold the following are equivalent

i) M is a complete Riemannian manifold

ii) M is a complete metric space with respect to the distance function $d(\cdot, \cdot)$.

iii) Every bounded subset of M (with respect to $d(\cdot, \cdot)$) is relatively compact

iv) For $x \in M$ and a curve C in $A_x M$ starting at the origin there is a curve τ in M starting from x that is developed on C.

4.2 BROWNIAN MOTION AND RELATED PROCESSES IN MANIFOLDS

The construction of Brownian motion and its horizontal lift to the bundle of orthonormal frames are sample path constructions that proceed from a sequence of piecewise smooth processes that converge to a Brownian motion in the tangent space of the initial value of the manifold-valued Brownian motion. Then it suffices to show that the sequence of manifold-valued processes obtained by these constructions converges uniformly to a process that is easily verified to be a manifold-valued Brownian motion or its horizontal lift.

For a piecewise C^1 curve $\tau = x_t,\ t \in [0,1]$ in a Riemannian manifold M there is a piecewise C^1 curve $\tilde{\tau} = C_t, t \in [0,1]$ in $A_{x_0}M$ such that $\tilde{\tau}$ is the development of τ into $A_{x_0}M$. Conversely given a piecewise C^1 curve in $A_{x_0}M$ starting at the origin there is a curve in M starting at x_0 that is developed on the curve in $A_{x_0}M$. These facts follow from the definition of the development and parallel displacement. This construction can be extended to give an isometry between the Sobolev space of $T_{x_0}M$-valued functions that are absolutely continuous and whose derivatives are square integrable and the Sobolev space of M-valued functions that are absolutely continuous and whose derivatives are square integrable. The construction of Brownian motion that is given here is an extension (almost surely) of this identification of Sobolev spaces. Since for many applications it is important to have the notion of parallelism defined for Brownian paths the (horizontal) lift of a Brownian motion in M is given in the bundle of orthonormal frames $O(M)$.

Theorem 2. Let $(B_t, t \in [0,1])$ be an n-dimensional standard T_aM-valued Brownian motion where M is a smooth, compact, connected, n-dimensional Riemannian manifold. For $u \in \pi^{-1}(a)$ where $\pi : O(M) \to M$ is the projection there is an $O(M)$-valued process $(F_t,\ t \in [0,1])$ with $F_0 \equiv u$ such that the process $(\pi(F_t),\ t \in [0,1])$ is an M-valued Brownian that is developed (almost surely) on $(B_t,\ t \in [0,1])$. The process $(F_t,\ t \in [0,1])$ is the (horizontal) lift of $(\pi(F_t),\ t \in [0,1])$ such that $F_0 \equiv u_0$ and $\pi(u) = a$.

While the proof of this result is not given here a brief sketch of the proof is described and the local stochastic differential equations are given.

Let $(t_i^{(n)},\ i = 0, 1, \ldots, 2^n)$ be the nth dyadic partition of $[0,1]$ and let $h(x) \in GL(n; \mathbb{R})$ be the coordinate transformation from the geodesic frames induced from u that are used in the local trivialization of $O(M)$ to the frames in the local trivialization of $L(M)$ given by the geodesic local coordinate system induced from u. Let $(C_t^{(n)},\ t \in [0,1])$ be the M-valued process whose development on T_aM is the process

$(B_t^{(n)},\ t \in [0,1])$ that is obtained from $(B_t,\ t \in [0,1])$ by linear interpolation using the points $(B_{t_i^{(n)}},\ i = 0, 1, \ldots, 2^n)$. Define

$$X_t^{(n)} = \exp_a^{-1}\left(C_{t\wedge T}^{(n)}\right) \tag{4.24}$$

where $\exp_a$ is the exponential map at a and T is the first fitting time of the boundary of the neighborhood of a that is used in the trivializations.

For $t \in \left[t_i^{(n)}, t_{i+1}^{(n)}\right]$ we obtain from Taylor's formula

$$\begin{aligned} X_t^{(n)} &= X_{t_i}^{(n)} + \frac{t-t_i}{t_{i+1}-t_i} h\left(X_{t_i}^{(n)}\right) g_{t_i}^{(n)} \left(B_{t_{i+1}} - B_{t_i}\right) \\ &\quad - \frac{1}{2}\Gamma\left(X_{t_i}^{(n)}\right)\left[h\left(X_{t_i}^{(n)}\right) g_{t_i}^{(n)} \frac{t-t_i}{t_{i+1}-t_i}\left(B_{t_{i+1}} - B_{t_i}\right)\right]^{(2)} \\ &\quad + \frac{1}{6} f\left(X_{t_i+\Theta(t_{i+1}-t_i)}\right)\left[h(X_{t_i}^{(n)}) g_{t_i}^{(n)} \frac{t-t_i}{t_{i+1}-t_i}(B_{t_{i+1}} - B_{t_i})\right]^{(3)} \end{aligned} \tag{4.25}$$

where $\Theta \in (0,1]$ and the superscript on the partition points has been suppressed for notational convenience.

In geodesic local coordinates let the connection form ω be given as

$$\omega = \sum \left(\alpha_{jk}^i dx^j\right) E_i^k$$

In the geodesic local coordinate system, the connection form for the geodesic frames evaluated for the process $(C_t^{(n)},\ t \in [0,1])$ is

$$\dot{g}_{t_s}^{(n)} g_{t_s}^{(n)-1} = -\sum \alpha_{jk}^i (X_{t_s}^{(n)}) \left[h(X_{t_s}^{(n)}) g_{t_s}^{(n)} \left(\frac{B_{t_{s+1}} - B_{t_s}}{t_{s+1} - t_s}\right)\right]^j E_i^k \tag{4.26}$$

For $t \in (t_s, t_{s+1})$

$$\begin{aligned} g_t^{(n)} &= g_{t_s}^{(n)} + \dot{g}_{t_s}^{(n)}(t - t_s) + \frac{1}{2}\ddot{g}_{t_s}^{(n)}(t-t_s)^{(2)} \\ &\quad + \frac{1}{6}\dddot{g}_{\,t_s+\beta(t-t_s)}(t-t_s)^{(3)} \end{aligned} \tag{4.27}$$

Rewriting (4.26) as

$$\dot{g}_t^{(n)} = A^{(n)}(t) g_t^{(n)} \tag{4.28}$$

and using this equation to compute the derivatives in (4.27) we have for $t \in (t_s, t_{s+1})$

$$g_t^{(n)} = g_{t_s}^{(n)} + A^{(n)}(t_s) g_{t_s}^{(n)}(t - t_s)$$

$$
\begin{aligned}
&+\frac{1}{2}A^{(n)}(t_s)A^{(n)}(t_s)g_{t_s}^{(n)}(t-t_s)^2 \\
&-\frac{1}{2}\sum\left(D\alpha^i_{jk}(X_{t_s}^{(n)})h(X_{t_s}^{(n)})\frac{t-t_s}{t_{s+1}-t_s}\left(B_{t_{s+1}}-B_{t_s}\right)\right) \\
&\left[h(X_{t_s}^{(n)})g_{t_s}^{(n)}\frac{t-t_s}{t_{s+1}-t_s}(B_{t_{s+1}}-B_{t_s})\right]^j E_i^k g_{t_s}^{(n)} \\
&+\frac{1}{2}\sum\alpha^i_{jk}(X_{t_s}^{(n)})\Gamma^i_{lm}(X_{t_s}^{(n)})\left[h(X_{t_s}^{(n)})g_{t_s}^{(n)}\frac{t-t_s}{t_{s+1}-t_s}(B_{t_{s+1}}-B_{t_s})\right]^l \\
&\left[h(X_{t_s}^{(n)})g_{t_s}^{(n)}\frac{t-t_s}{t_{s+1}-t_s}(B_{t_{s+1}}-B_{t_s})\right]^m E_i^k \\
&+\frac{1}{6}\sum h_i^j\left(X_{t_s+\Theta(t-t_s)}^{(n)}, g_{t_s+\Theta(t-t_s)}^{(n)}\right) \\
&\left[\frac{t-t_s}{t_{s+1}-t_s}\left(B_{t_{s+1}}-B_{t_s}\right)\right]^{(3)} F_j^i
\end{aligned}
\tag{4.29}
$$

where $\Theta \in [0,1], (h_j^i,\ i,j=1,\ldots,n)$ are trilinear continuous functions and $(F_j^i,\ i,j=1,\ldots,n)$ is a basis for $gl(n)$.

The local stochastic differential equations that must be satisfied for the construction of the M-valued Brownian motion and its lift can be obtained by formally passing to the limit in (4.25) and (4.29). This passage to the limit has been justified in the proof of Theorem 2. Thus we have

$$
dX_t = h(X_t)g_t dt - \frac{1}{2}tr(h^T(X_t)\Gamma(X_t)h(X_t))dt \tag{4.30}
$$

$$
\begin{aligned}
dg_t = &-\sum\alpha^i_{jk}(X_t)[h(X_t)g_t dB_t]^j E_i^k g_t \\
&+\frac{1}{2}\sum(h^T(X_t)\alpha^i_{\cdot k}(X_t), h^T(X_t)\alpha^l_{\cdot m}(X_t))E_i^k E_l^m g_t dt \\
&-\frac{1}{2}\sum tr(h^T(X_t)D\alpha^i_{\cdot k}(X_t)h(X_t))E_i^k g_t dt \\
&+\frac{1}{2}\sum\alpha^i_{\cdot k}(X_t)tr(h^T(X_t)\Gamma^i(X_t)h(X_t))E_i^k g_t dt
\end{aligned}
\tag{4.31}
$$

where $X_0 \equiv 0,\ g_0 = I,\ tr(h^T\Gamma h)$ is a vector of traces of the matrices where Γ is considered as a matrix in the two lower indices and $(\alpha^i_{\cdot k})$ is a vector formed by suppressing the third index.

The Laplacian Δ on M acting on a smooth function f is

$$
(\Delta f)(x) = \sum g^{ij}(x)\frac{\partial^2 f}{\partial x^i \partial x^j} - \sum g^{ij}(x)\Gamma^k_{ij}\frac{\partial f}{\partial x^k} \tag{4.32}
$$

where (g^{ij}) is the inverse of (g_{ij}). A local stochastic differential equation for the diffusion with the Laplacian as infinitesimal generator is (4.30) for $(X_t,\ t \in [0,1])$.

The four terms in the stochastic differential equation (4.31) for $(g_t,\ t \in [0,1])$ occur as follows:

i) the first term on the right hand side of (4.31) is the usual term for piecewise smooth curves

ii) the second term occurs from the quadratic oscillation of Brownian motion

iii) the third term occurs because α is a function of the position

iv) the fourth term occurs by the differentiation of the local coordinates.

While it is not true that a Brownian motion in an arbitrary complete Riemannian manifold does not have a finite escape time (explosion), this statement is true for many noncompact, complete manifolds. To verify this statement Brownian motion in a manifold is compared with Brownian motion in a noncompact, rank one symmetric space, that is, a space with constant negative sectional curvature.

Let R be the curvature that is defined as

$$\begin{aligned} R(X,Y)Z &= [\nabla_X, \nabla_Y]Z - \nabla_{[X,Y]}Z \\ &= u(2\Omega(X^*,Y^*))(u^{-1}Z) \end{aligned} \tag{4.33}$$

for $X,Y,Z \in T_xM$ The Ricci tensor field S is the covariant tensor field of degree 2 that is defined for an orthonormal basis $(V_1,\ldots,V_n)$ of T_xM as

$$\begin{aligned} S(X,Y) &= \sum_{i=1}^{n} g(R(V_i,X)Y,V_i) \\ &= \sum_{i=1}^{n} R(V_i,Y,V_i,X) \end{aligned} \tag{4.34}$$

for $X,Y \in T_xM$. It is easy to verify that $S(X,Y) = S(Y,X)$.

A useful result for the nonexistence of explosions is the following:

Theorem 3. Let M be a complete Riemannian manifold with Ricci curvature bounded below by a constant. Then the M-valued Brownian motion does not have explosions, that is,

$$P(t,x,M) = 1 \tag{4.35}$$

for all $x \in M$.

While the subsequent applications are usually restricted to compact Riemannian manifolds, many of the results can be easily extended to noncompact, complete Riemannian manifolds whose Ricci curvature is bounded below by a constant.

For the development of a stochastic calculus (for Brownian motion) in a manifold the first topic is the construction of stochastic integrals. The process $(X_t, t \geq 0)$ is an M-valued continuous Brownian local martingale if and only if for each path and some local coordinate system

$$dX_t^i + \frac{1}{2}\sum \Gamma_{jk}^i(X_t)d\langle X^j, X^k\rangle_t = dY_t^i \tag{4.36}$$

where dY_t^i, $i = 1, \ldots, n$ are the differentials of n real-valued Brownian martingales. The equation (4.36) follows immediately from (4.30).

The symmetric or Fisk-Stratonovich integral of a one-form α with respect to a (Brownian) semimartingale is

$$I_S(\alpha) = \sum_{i,j} \int a_i(X_s)dX_s^i + \frac{1}{2}\int D_i a_j(X_s)d\langle X^i, X^j\rangle_s \tag{4.37}$$

where $\alpha = \sum a_i dx^i$, the first integral is an Itô integral and $D_i = \frac{\partial}{\partial x^i}$. It is easy to verify that this definition is intrinsic. The advantage of this symmetric integral is that it transforms in the usual fashion under mappings. If I_I is the Itô integral then

$$\begin{aligned} I_I(\alpha) - I_S(\alpha) = \frac{1}{2}\sum \int & [a_k(X_s)\Gamma_{ij}^k(X_s) \\ & - D_i a_j(X_s)]\, d\langle X^i, X^j\rangle_s \end{aligned} \tag{4.38}$$

If the integrator is Brownian motion then

$$I_I(\alpha) - I_S(\alpha) = \frac{1}{2}\int \delta\, \alpha(X_s)ds \tag{4.39}$$

where δ is the adjoint of the exterior derivative d so that

$$\delta\alpha(x) = \sum g^{ij}(x)(D_i a_j(x) - \sum \Gamma_{ij}^m a_m(x)) \tag{4.40}$$

The equation (4.39) is easily verified using the well known relation between these integrals in Euclidean space and parallel translation along the paths of the M-valued Brownian motion. Translate a local frame field along the paths of Brownian motion to the initial point of Brownian motion. Compute the relation between the two integrals in the initial tangent space in the family of frame fields that have been parallel translated and parallel translate these back along the paths of Brownian motion to obtain (4.39). This result follows from the fact that an observer relating the two definitions of integrals in an orthonormal frame along the Brownian paths performs the same computation as in Euclidean space so he can detect no difference between these two computations.

Stochastic integrals can also be constructed with values in a compact Riemannian manifold from a continuous martingale integrator (e.g. Brownian motion) that is given in the tangent space of the initial point of the stochastic integral and from a stochastic tensor field of linear endomorphisms of the tangent bundle. The integrals that are formed are continuous processes that suitably preserve the martingale property. This construction is a generalization of the development of Brownian motion in a Riemannian manifold.

4.3 THREE APPLICATIONS TO DIFFERENTIAL GEOMETRY

Initially some applications to differential geometry are made where the stochastic calculus for a manifold-valued Brownian motion plays a central role. The first two applications use a generalization of a formula of M. Kac.

Let $\frac{\partial}{\partial t} - L - B$ be a differential operator where L is a second order elliptic operator and B is a linear transformation. If $(X_t, t \geq 0)$ is a diffusion with infinitesimal generator L and $(Y_t, t \geq 0)$ satisfies

$$\frac{dY}{dt} = Y(t)\,B(t, X_t) \qquad (4.41)$$
$$Y(0) = I$$

then $u(t,x) = E_x[Y(t)\varphi(X_t)]$ satisfies $\frac{\partial}{\partial t} - L - B = 0$ with initial data φ. The Laplacian on p-forms has the local expression

$$\Delta = \sum g^{ij} \nabla_j \circ \nabla_i + D^p R \qquad (4.42)$$

where (g^{ij}) is the inverse of the Riemannian metric $(g_{ij}), R$ is the curvature tensor field of type (2, 2) and $D^p R$ is the derivation extension of R to p-forms.

A useful identity in linear algebra is the following. Let $A_1, \ldots, A_k \in$ Hom (V, V) where $k \in \{0, \ldots, d\}$ and $d = \dim V$. Then

$$\sum (-1)^p Tr(D^p A_1 \circ \ldots \circ D^p A_k) = \begin{cases} 0 & \text{if } k < d \\ (-1)^d K & \text{if } k = d \end{cases} \qquad (4.43)$$

where K is the coefficient of $x_1, \ldots, x_k$ in $\det(x_1 A_1 + \ldots, +x_k A_k)$.

A vanishing theorem of Bochner for a compact, connected Riemannian manifold M with positive Ricci curvature states that the first cohomology group of M is trivial or equivalently that there are no nontrivial harmonic 1-forms. Since $D^1 R = -S$ it easily follows from the formula of M. Kac with $L = \Delta$ that for a complete connected Riemannian manifold M with nonnegative Ricci curvature that is positive at some point the first cohomology group of M with compact support is trivial. It is also possible to allow the Ricci curvature to be negative in some regions and still obtain the same vanishing theorem.

Many topological invariants of manifolds can be described as geometric computations by index theorems. Probably the simplest, general example of this phenomena is the Euler-Poincaré characteristic.

Theorem 4. Let M be a smooth, compact oriented Riemannian manifold of dimension $2n$. Let $e^p(t, x, y)$ be the fundamental solution of the heat operator $\frac{\partial}{\partial t} - \Delta$ acting on the p-forms for $p \in \{0, 1, \ldots, 2n\}$.

Let

$$\operatorname{Tr} e = \sum_{p=0}^{2n} (-1)^p \operatorname{Tr} e^p \tag{4.44}$$

Then

$$\lim_{t \downarrow 0} \operatorname{Tr} e = C \tag{4.45}$$

where C is the Chern polynomial whose integral over M is the Euler-Poincaré characteristic.

An elementary proof of this result can be given using the Kac formula (4.41) and the equation (4.43).

Affine Lie algebras are a family of infinite dimensional Lie algebras that have structural properties that are analogous to semisimple Lie algebras. Affine Lie algeras provide a natural setting to describe some theta functions. The theta function, θ, associated with the compact simple Lie algebra g is

$$\theta(k, b) = e^{-2\pi i b|\rho|^2} \sigma(k) \Pi_{n=1}^{\infty} \det\left(1 - e^{-2\pi i b n} Ad(e^k)\right) \tag{4.46}$$

where $|\cdot|^2$ is the quadratic form induced from the Killing form, $\operatorname{Im} b < 0$, $k \in h$ (a Cartan subalgebra of g), Ad is the adjoint representation of G, ρ is one half of the sum of the positive roots, and

$$\sigma(k) = \Pi_{\alpha\in R_+}\left(e^{\frac{(\alpha,k)}{2}} - e^{\frac{-(\alpha,k)}{2}}\right) \tag{4.47}$$

where R_+ is the family of positive roots.

Brownian motion in G, the simply connected compact semisimple Lie group with Lie algebra g, can be used to describe the theta function $\theta(k,b)$ where $ib \in \mathbb{R}_+$. In addition it can be used to obtain some other descriptions of a theta function.

Theorem 4.5 For the theta function (4.46) the following equality is satisfied

$$\theta(k,b) = e^{-2\pi ib|\rho|^2}\sigma(k)\int d\mu_k\left(\cdot|4\pi ib, O_{e^{4\pi x_\rho}}\right) \tag{4.48}$$

where $\mu_k\left(\cdot|4\pi ib, O_{e^{4x_\rho}}\right)$ is the Wiener measure on the Borel subsets of $C([0,1],G)$ with initial value e^k and variance parameter $4\pi ib$ that is pinned at time 1 to the orbit $O_{e^{4\pi x_\rho}}$ formed by conjugation that contains $e^{4\pi x_\rho}$ and x_ρ is the dual of ρ.

4.4 TWO APPLICATIONS TO PROBABILITY

Two applications to probability are given that use the stochastic calculus for Brownian motion in a manifold. The first application is the question of absolute continuity for Wiener measure in a manifold and the second application is to a function space law of the iterated logarithm for a manifold-valued Brownian motion. The absolute continuity result is important in applications of stochastic calculus in manifolds to stochastic control and filtering.

Theorem 4.2 shows that the development and its inverse can be extended almost surely to the paths of Brownian motion. If P is the Wiener measure on T_aM-valued continuous functions and $\tilde{P}$ is the Wiener measure on M-valued continuous functions, there is an isometry between $L^2(P)$ and $L^2(\tilde{P})$ and thus neglecting null sets there is a bijection between the σ-algebras for the two Brownian motions. With this identification many results for Brownian motion in Euclidean space have analogues for Brownian motion a manifold.

One example of this pheonomenon which is important in applications is the transformation of Wiener measure by a Radon-Nikodym derivative.

Theorem 4.6 Let $(\varphi_t,\ t \in [0,1])$ be a TM-valued predictable process with respect to the M-valued Brownian motion $(C_t,\ t \in [0,1])$ such that

$$\int_0^1 \langle \varphi_s, \varphi_s \rangle_{C_s}\, ds < \infty \qquad \text{a.s. } \tilde{P} \tag{4.49}$$

where $\langle \cdot, \cdot \rangle$ is the Riemannian metric on M. Then $(M_t, \tilde{\mathcal{F}}_t, \tilde{P})$ is a continuous local martingale where $(\tilde{\mathcal{F}}_t)$ is the completion of the natural family of σ-algebras for $(C_t,\ t \in [0,1])$ and

$$M_t = \exp\left[\int_0^t \langle \varphi_s, dC_s \rangle_{C_s} - \frac{1}{2}\int_0^t \langle \varphi_s, \varphi_s \rangle ds\right] \tag{4.50}$$

If $(\varphi_t,\ t \in [0,1])$ is uniformly bounded then $(M_t, \tilde{\mathcal{F}}_t, \tilde{P})$ is a continuous martingale and

$$\Theta(dC_t) = \varphi_t dt + d\tilde{B}_t \tag{4.51}$$

where Θ is the canonical 1-form and $(d\tilde{B}_t)$ are the formal vectors of a Brownian motion with respect to $(\tilde{\mathcal{F}}_t, \bar{P})$ where $d\bar{P} = M d\tilde{P}$.

The classical law of the iterated logarithm for a real-valued standard Brownian motion $(B_t,\ t \geq 0)$ states that the set of cluster points of the family of random variables

$$\frac{B(t)}{(2t \log \log t)^{1/2}}$$

as $t \longrightarrow \infty$ is almost surely $[-1, +1]$. The function space version of this law states that the set of cluster points of $(C_n(t),\ t \in [0,1]\, n \geq 3)$ in the norm topology is almost surely the unit ball of $L_0^{2,1}(\mathbb{R})$, the Sobolev space of absolutely continuous functions on $[0,1]$ whose derivatives are square integrable.

A manifold version of this latter result is given now.

Theorem 4.7 Let M be a connected, compact, Riemannian manifold of dimension N and let $(B_t,\ t \in [0,1])$ be an N-dimensional T_aM-valued standard Brownian motion. For $n \geq 3$ let $(C_n(t),\ t \in [0,1])$ be the M-valued process that is developed on

$$\frac{B(nt)}{(2n \log \log n)^{1/2}}$$

Then the family of M-valued processes $(C_n(t),\ t \in [0,1],\ n \geq 3)$ cluster almost surely in the uniform topology on $C_a([0,1], M)$ at every point of the unit ball of $L_a^{2,1}(M)$, the Sobolev space of absolutely continuous M-valued functions on $[0,1]$ with initial value $a \in M$ whose derivatives are square integrable using the Riemannian metric.

4.5 SOME APPLICATIONS TO STOCHASTIC CONTROL AND FILTERING

Some applications of stochastic calculus in manifolds are described for the control of stochastic systems in manifolds. The first application is an existence result for optimal controls of completely observed stochastic systems in a compact, connected Riemannian manifold. The control appears in the vector fields that appear as the drift term in the local stochastic differential equation. These vector fields satisfy a convexity property in the control. In this sense the result is analogous to some well known existence results in deterministic optimal control.

Using the canonical 1-form on the compact, connected Riemannian manifold M the stochastic system is given as

$$\Theta(dX_t) = f(t, X, u(t, X))dt + dB_t \tag{4.52}$$

where (dB_t) are the formal vectors of an M-valued Brownian motion. Let $(\mathcal{F}_t,\ t \in [0,1])$ be the completion of the family of sub-σ-algebras generated by the M-valued Brownian motion in $[0,1]$ starting at $a \in M$. Let the controls be chosen from U, a smooth, compact manifold and denote its Borel σ-algebra as $\mathcal{B}_U$. The following conditions are imposed on the drift vector fields f

i) $f(t, \cdot, \cdot)$ is $\mathcal{F}_t \otimes \mathcal{B}_U$ measurable

ii) $f(t, z, \cdot) : u \rightarrow T_{z(t)}M$ is continuous for all (t, z)

iii) The family $(f(t, z, u),\ t \in [0,1],\ z \in C_a([0,1], M),\ u \in \mathcal{U})$ is uniformly bounded

iv) For each $(t, z) \in [0,1] \times C_a([0,1], M)$ the set $f(t, z, U) = \{f(t, z, u) : u \in U\}$ is a closed convex subset of $T_{z_t}M$.

An admissible control is a map $u : [0,1] \times C_a([0,1], M) \longrightarrow U$ that is $\mathcal{B} \otimes F$ measurable such that u_t is adapted to $\mathcal{F}_t$ for all $t \in [0,1]$. The solution to the stochastic system (4.52) is defined by a measure on $(\Omega, \mathcal{F})$ by a Radon-Nikodym derivative using Theorem 4.6. For the drifts f as given above it can be shown that the associated family of Radon-Nikodym derivatures that are formed by using Theorem 4.6 is a closed, convex, uniformly integrable subset of $L^1(P)$.

Theorem 4.8 Let $L : C_a([0,1], M) \to \mathbb{R}$ be a bounded $\mathcal{F}$-measurable function and define the cost associated with $u \in \mathcal{U}$, the family of admissible controls, as

$$J(u) = E[L\rho(g_u)] \tag{4.53}$$

where g_u is the drift induced from f by u and ρ is the Radon-Nikodym derivative. Then there is an optimal control $u^* \in U$ such that

$$J(u^*) \leq J(u) \tag{4.54}$$

for all $u \in \mathcal{U}$.

For partially observed stochastic systems in manifolds one can obtain necessary and sufficient conditions for optimal stochastic controls by Hamilton-Jacobi or dynamic programming methods. In this case there is a stochastic vector field that is along the sample paths of the observation process that appears in the equation for optimality.

In addition to stochastic control problems that can be formulated in manifolds one can also formulate stochastic filtering problems in manifolds. This fact should not be surprising because the stochastic control of a partially observed stochastic system usually requires an application of filtering. One simple generalization of the nonlinear filtering problem in Euclidean spaces is to a nonlinear filtering problem where the observations occur in a manifold and the state evolves in a Euclidean space.

Let $(X_t,\, t \in [0,1])$ be a continuous $\mathbb{R}^n$-valued stochastic process and let M be a compact, connected Riemannian manifold. Let $f : [0,1] \times \mathbb{R}^n \times M \to TM$ be a continuous vector field such that $f(\cdot,\cdot,x) \in T_xM$ and f is differentiable in its third variable with a uniformly bounded derivature. Consider the observation stochastic differential equation

$$\Theta(dY_t) = f(t, X_t, Y_t)dt + dB_t \tag{4.55}$$

where Θ is the canonical 1-form and (dB_t) are the formal vectors of an M-valued standard Brownian motion that is independent of $(X_t,\, t \in [0,1])$. It is easy to construct the solution $(Y_t,\, t \in [0,1])$ of (4.55). Let $\mathcal{G}_t$ be the completion of

$\sigma(Y_u,\ u \in [0,t])$, the sub-$\sigma$-algebra generated by $(Y_u,\ u \in [0,t])$. It can be shown that $(Y_t,\ t \in [0,1])$ satisfies the stochastic differential equation

$$\Theta(dY_t) = \hat{f}_t dt + dC_t \tag{4.56}$$

where $\hat{f}_t = E[f(t, X_t, Y_t)|\mathcal{G}_t]$ and (dC_t) are the formal vectors of a Brownian motion adapted to $(\mathcal{G}_t,\ t \in [0,1])$. The conditional expectation $\hat{f}_t$ is easily defined by using 1-forms in T^*M. This result plays a basic role in the nonlinear filtering solution as it does for processes in Euclidean space.

Theorem 4.9 If $g : \mathbb{R}^n \to \mathbb{R}$ is a bounded, twice continuously differentiable function where the first two derivatives are bounded and $(X_t,\ t \in [0,1])$ is a diffusion process with infinitesimal generator $\mathcal{L}$, then $(E[g(X_t)|\mathcal{G}_t],\ t \in [0,1])$ satisfies the stochastic equation

$$\begin{aligned} dE[g(X_t)|\mathcal{G}_t] &= E[\mathcal{L}g_t|\mathcal{G}_t]dt \\ &\quad + \langle E[g_t f_t|\mathcal{G}_t] - E[g_t|\mathcal{G}_t]\hat{f}_t,\ dC_t\rangle \end{aligned} \tag{4.57}$$

where $g_t = g(X_t)$, $f_t = f(t, X_t, Y_t)$, $\hat{f}_t = E[f(t, X_t, Y_t)|\mathcal{G}_t]$, $(C_t,\ t \in [0,1])$ is defined in (4.56) and $\langle \cdot, \cdot \rangle$ is the Riemannian metric.

In addition to stochastic processes with continuous sample paths it is natural to investigate some discontinuous processes in manifolds. The most natural generalization of discontinuous processes to manifolds are manifolds that are vector bundles and the discontinuities of the discontinuous process occur in the fibres of the vector bundle. In this case the differential structure is still important. For example consider the tangent bundle of a Riemannian manifold. Physically one can imagine that this is a mathematical model for a particle that has collisions so that its velocity is randomly changed. If the discontinuities of the discontinuous process occur in the fibres of the vector bundle then some of the techniques for the analysis of discontinuous processes in Euclidean spaces can be utilized even though the Euclidean space is varying in the fibres of the vector bundle. Optimal control problems can be posed and solved. For example for such a partially observed stochastic system a stochastic equation can be obtained that characterizes an optimal control. Furthermore it is possible to solve filtering problems for such discontinuous processes. For example, if the observations include the process in the base manifold and the times of discontinuity in the fibres of the vector bundle, then a filtering result is naturally obtained.

4.6 SOME SOLVABLE STOCHASTIC CONTROL PROBLEMS IN MANIFOLDS

Some examples are presented now of stochastic control problems in manifolds where explicit solutions can be found. The examples include both compact and noncompact Riemannian manifolds. Even in Euclidean spaces there are relatively few examples of explicitly solvable stochastic control problems.

The first example is a stochastic system in real hyperbolic n-space $\mathbb{H}^n(\mathbb{R})$ where $n \geq 3$. The stochastic system is Brownian motion plus a control in the drift. A geometric model for $\mathbb{H}^n(\mathbb{R})$ is the open unit ball $\mathbb{R}^n$. If the unit ball is given the Riemannian structure

$$ds^2 = 4(1-|y|^2)^{-2}(dy_1^2 + dy_2^2 + \ldots + dy_n^2)$$

then the manifold has constant sectional curvature -1. Manifolds of constant negative sectional curvature are globally trivial and a useful global coordinate system at the origin is geodesic polar coordinates defined by the map

$$\mathrm{Exp}_0 Y \longmapsto (r, \theta_1, \theta_2, \ldots, \theta_{n-1}) \tag{4.58}$$

where $Y \in T_0\,\mathbb{H}^n$, $r = |Y|$ with $|\cdot|$ the Riemannian metric and $(\theta_1, \theta_2, \ldots, \theta_{n-1})$ are the local coordinates of the unit vector $Y/|Y|$. In these coordinates the Laplace-Beltrami operator $\Delta_{\mathbb{H}^n}$ is

$$\Delta_{\mathbb{H}^n} = \frac{\partial^2}{\partial r^2} + (n-1)\coth r \frac{\partial}{\partial r} + (\sinh r)^{-2}\Delta_{S^{n-1}} \tag{4.59}$$

where $\Delta_{S^{n-1}}$ is the Laplacian on the unit sphere in $T_0\,\mathbb{H}^n$.

Let $(Z_t,\, t \geq 0)$ be the controlled diffusion in $\mathbb{H}^n(\mathbb{R})$ with the infinitesimal generator expressed in geodesic polar coordinates at the origin as

$$\begin{aligned}\frac{1}{2}\Delta_{\mathbb{H}^n} + u\frac{\partial}{\partial r} = \frac{1}{2}\frac{\partial^2}{\partial r^2} + \frac{1}{2}(n-1)\coth r\frac{\partial}{\partial r} + \\ + u\frac{\partial}{\partial r} + \frac{1}{2}(\sinh r)^{-2}\Delta_{S^{n-1}}\end{aligned} \tag{4.60}$$

for $r > 0$. Controls could also be assumed in the $(\theta_1, \ldots, \theta_{n-1})$ coordinates so that the stochastic system is Brownian motion plus a control in the drift term. However it will become apparent that these additional controls are superfluous. For the analysis of the controlled diffusion it suffices to describe the stochastic differential equation for the radial part of the diffusion. This equation is

$$dX(t) = \left(\frac{n-1}{2}\coth X(t) + U(t)\right) dt + dB(t) \tag{4.61}$$

where $(B_t,\, t \geq 0)$ is a real-valued standard Brownian motion and $X(0) = \alpha > 0$.

Let $T_1 > 0$ be chosen such that

$$\sup_{0 \leq t \leq T_1} g(t) \leq \frac{n-1}{2} \tag{4.62}$$

where g is the unique positive solution of the Riccati differential equation

$$g' + \frac{n-1}{2}g - \frac{1}{4}g^2 + 1 = 0 \tag{4.63}$$

$$g(T_1) = 0 \tag{4.64}$$

An admissible control at time t is a measurable function of the state $Z(t)$ that is a smooth function on $\mathbb{H}^n \backslash \{O\}$ such that the solution of (4.61) exists and is unique in a sample path sense.

Let $T \in (0, T_1]$ be fixed. Define the cost functional in the geodesic polar coordinates at the origin to be

$$J(U) = E_\alpha \int_0^T \left[\cosh X(t) + \frac{\sinh^2 X(t)}{\cosh X(t)} U^2(t)\right] dt \tag{4.65}$$

where $X(0) = \alpha > 0$. It is clear that there are admissible controls that give a finite cost, e.g. $U \equiv 0$.

Theorem 4.10 Fix $n \geq 3$. The stochastic control problem in $\mathbb{H}^n(\mathbb{R})$ described by (4.60) and (4.65) has an optimal control u^* that in geodesic polar coordinates at O is

$$u^*(s, r) = -\frac{1}{2}g(s)\cosh r \tag{4.66}$$

where $s \in [0, T]$, $r > 0$ and g is the unique positive solution of (4.63) with $g(T) = 0$.

An example of a stochastic optimal control problem in a sphere of dimension at least three is described and explicitly solved. The spheres are the compact analogues of the real hyperbolic spaces and this stochastic control problem in spheres is solved by finding a smooth solution to the Hamilton-Jacobi equation as was done for the stochastic control problem in the real hyperbolic spaces. The controlled diffusion in a sphere is Brownian motion plus a control in the drift. The n-dimensional sphere S^n is diffeomorphic to the rank one symmetric space $SO(n+1)/SO(n)$. For $n \geq 2$ these spaces are the simply connected, compact Riemannian manifolds of constant positive sectional curvature. The Killing form metric on S^n is used. It is obtained by multiplying the usual Riemannian metric (with sectional curvature + 1) by $2n$. The maximal distance between any two points of S^n using the Killing form metric is L where

$$L = \pi\sqrt{\frac{p}{2}} \tag{4.67}$$

$$p = n - 1 \tag{4.68}$$

Choose an origin $O \in S^n$. The antipodal point of O is the submanifold A_0 that is the distance L from O. The mapping $\mathrm{Exp}_0 : T_0 S^n \to S^n$ is a diffeomorphism of the ball $B_L(0) = \{x \in T_0 S^n : |x| < L\}$ onto the open set $S^n \backslash A_0$. Geodesic polar coordinates are used.

Let $(Z_t,\, t \geq 0)$ be the controlled diffusion in S^n with the infinitesimal generator expressed in geodesic polar coordinates at the origin as

$$\frac{1}{2}\Delta_{S^n} + u\frac{\partial}{\partial r} = \frac{1}{2}\frac{\partial^2}{\partial r^2} + \frac{1}{2}\lambda p \cot \lambda r \frac{\partial}{\partial r} + u\frac{\partial}{\partial r} + \frac{1}{2}\Delta_{S_r} \tag{4.69}$$

where $r \in (0), L)$, $\lambda = (2p)^{\frac{-1}{2}}$ and Δ_{S_r} is the Laplace-Beltrami operator on the sphere of radius r from the origin. The local stochastic differential equation for the radial part of the diffusion is

$$dX(t) = \left(\frac{1}{2}\lambda p \cot \lambda X(t) + U(t)\right) dt + dB(t) \tag{4.70}$$

where $(B(t),\, t \geq 0)$ is a real-valued standard Brownian motion and $X(0) = \alpha \in (0, L)$ and $X(t) \in (0), L)$.

Let $T_1 > 0$ be chosen such that the Riccati differential equation

$$g' - \frac{1}{2}\lambda^2(1+p)g - \frac{1}{4}\lambda^2 g^2 - 1 = 0 \tag{4.71}$$

$$g(T_1) = 0 \tag{4.72}$$

has one and only one solution in $[0, T_1]$ and

$$\sup_{0 \leq t \leq T_1} |g(t)| < \frac{1}{2}$$

An admissible control at time t is a measurable function of $Z(t)$ that is a smooth function on $S^n \backslash \{O\}$ such that the solution of (4.70) exists and is unique in a sample path sense.

Let $T \in (0, T_1]$ be fixed. Define the cost functional in the geodesic polar coordinates at the origin O be

$$J(U) = E_\alpha \int_0^T \left[-\cos \lambda X(t) + \frac{\sin^2 \lambda X(t)}{\cos \lambda X(t)} U^2(t) \right] dt \tag{4.73}$$

where $X(0) = \alpha \in (0, L)$ and $\lambda = (2p)^{\frac{-1}{2}}$.

Theorem 4.11 Fix $n \geq 3$. The stochastic control problem in S^n described by (4.69) and (4.73) has an optimal control U^* that in geodesic polar coordinates at the origin is

$$U^*(s, r) = \frac{1}{2} g(s) \lambda \cot \lambda r \tag{4.74}$$

where $s \in [0, T]$, $r \in (0, L)$ and g is the solution of the Riccati differential equation (4.71) with $g(T) = 0$.

4.7 A SOLVABLE ESTIMATION PROBLEM IN MANIFOLDS

Lie groups and symmetric spaces provide sufficient structure that allows one to make explicit computations in many problems. To provide another example of this phenomenon consider real hyperbolic three space $\mathbb{H}^3(\mathbb{R})$. The problem is to find an explicit expression for the conditional probability density of the initial condition of an $\mathbb{H}^3(\mathbb{R})$-valued Brownian motion given a scalar-valued observation of the Brownian motion at some positive time.

Theorem 4.12 Let $sl(2, \mathbb{C}) = k + p$ be a Cartan decomposition and $a \subset p$ be a maximal abelian subspace. Fix a Weyl chamber $a^+ \subset a$ and let $A^+ = \exp(a^+)$. Let $(Y_t, t \geq 0)$ be a standard $\mathbb{H}^3(\mathbb{R})$-valued Brownian motion with the initial value $\exp(L)$ where L is a random variable in the Weyl chamber a^+ with probability density function p_L. The random variable L is independent of the Brownian motion $(Y_t Y_0^{-1}, t \geq 0)$. Let f be a smooth one-to-one radial function in the unit ball model of $\mathbb{H}^3(\mathbb{R})$ and for fixed $t_0 > 0$ let $Z = f(|Y_{t_0}|)$. The conditional probability density of L given Z, $p_{L|Z}$, is

$$p_{L|Z}(l|z) = \frac{p_{LZ}(l,z)}{p_Z(z)} \tag{4.75}$$

where

$$\begin{aligned} p_{LZ}(l,z) = c\, p_L(l) \sinh^{-1} l \sinh^{-2} r [e^l q(r, t_0 + l; t_0) \\ + q(r, -t_0 - l; t_0)) - e^{-l}(q(r, l - t_0; t_0) \\ - q(r, t_0 - l; t_0))] \end{aligned} \tag{4.76}$$

where $r = f^{-1}(z)$, c is a constant, $q(\cdot,\cdot;t)$ is the Gaussian probability density function on $\mathbb{R}^1$ with variance parameter t and $p_Z(z) = E[p_{LZ}(L,z)]$.

Notes

§2. For more information on differential geometry the well known books [39, 47, 57] can be consulted. The frame bundle or moving frames approach to differential geometry was initiated by É. Cartan (e.g. [7]). The development of a smooth curve in a manifold was defined by É. Cartan. Much of the differential geometry that was presented here is associated with moving frames because it provides a global approach to many questions in differential geometry. This global approach is especially suited for stochastic calculus in manifolds.

§3. The local description of a diffusion process in a manifold in terms of Brownian motion and a stochastic differential equation was initiated by K. Itô [41-42]. Subsequently this local description was extended by others (e.g. [29, 34]). The notion of parallelism along the sample paths of Brownian motion was introduced by Itô [43] in the special case of flat manifolds. Dynkin [29] defined parallelism for tensors along Brownian paths. In the early seventies there was renewed interest in Brownian motion in manifolds. Eels and Elworthy [30] approached the construction of Brownian motion in a manifold using functional analysis and Hilbert and Banach manifolds. Subsequently they have adopted a sample path approach [31]. The author [10, 12, 14] provided an independent, sample path construction of Brownian motion and its horizontal lift in manifolds for use in the control and the filtering of stochastic systems in manifolds. The construction that is outlined here is taken from [12]. Malliavin [49-50] had studied these stochastic constructions in his extensions of the vanishing theorem of Bochner [65]. Pinsky [59-60] provided another construction of Brownian motion in a manifold. Subsequently there were other contributors to the construction of Brownian motion and related processes in manifolds. Real-valued (Itô) stochastic integrals for a manifold-valued Brownian motion were constructed and used extensively by the author [10, 12, 14-17, 19-20] in the analysis of stochastic control and filtering. The general theory was developed by Meyer [54]. The relation between the Itô integral and the symmetric integral (39) for Brownian motion is easily obtained from the corresponding result in Euclidean space and parallelism along Brownian paths [21]. The construction of some manifold-valued stochastic integrals was given by the author [13].

The sufficient condition that is described for a manifold-valued Brownian motion not to have an explosion in finite time was given by Yau [64]. Initially it had been conjectured that for every complete Riemannian manifold the associated Brownian motion does not have explosions. However this is not true [9].

§4. The well known formula of M. Kac is given in [45] and its generalizations are given in [2, 62]. The stochastic generalization of the vanishing theorem of Bochner [65] was given by Malliavin [49]. An analytic treatment was given by Yau [63]. Subsequently other generalizations have been made.

The stochastic approach to the local formulae of various index theorems was given by the author [18]. Subsequently it was rediscovered [5] and there have been a number of papers on this topic using the stochastic approach (e.g. [35]). Some other approaches to the local formulae of index theorems used another construction of the fundamental solutions or invariance theory (e.g. [1, 36, 58]). In both of these approaches the proofs are more complicated than the proofs using the stochastic approach. The stochastic approach to affine Lie algebras was initiated by Frenkel [33] and the author [22] used stochastic methods to obtain the various identities [48, 51] for a theta function associated with a compact semisimple Lie algebra. Affine Lie algebras are an important subset of Kac Moody algebras [46, 55]. Various other geometric applications of stochastic calculus exist.

§5. The study of the absolute continuity for Wiener measure in Euclidean space was initiated by Cameron and Martin [6]. The result given here is analogous to a result of Girsanov [37] for Brownian motion in Euclidean space. The absolute continuity result here was obtained by the author [10, 12, 14] for his use in stochastic control and filtering. Subsequently it was rediscovered in [40] and expressed in terms of the Fisk-Stratonovich stochastic integral. The function space law of the iterated logarithm for a manifold-valued Brownian motion was given by the author [11].

§6. The applications of stochastic calculus in manifolds to stochastic control and filtering were initiated and developed by the author in a series of papers [10, 12, 14-17, 19-21]. The result given on the existence of optimal controls for completely observed stochastic systems is given in [10, 12]. The latter paper was accepted for publication in March 1976 but the journal backlog of papers delayed its publication until 1979. A necessary and sufficient condition for an optimal control of a partially observed stochastic system was given in [14]. Recently these methods have been refined for completely observed stochastic systems [8]. The filtering result described here is given in [15]. Subsequently this result has been generalized and refined [20, 56, 61]. Some results for the stochastic control and filtering of discontinuous processes in a vector bundle have been given by the author [17, 19]. These papers are apparently the first ones that consider discontinuous processes in a vector bundle and develop the stochastic calculus for these processes.

§7. The examples of solvable stochastic control problems in manifolds given here were obtained by the author [24, 28]. The example of a real hyperbolic space is apparently the first example of a solvable stochastic control problem in a noncompact manifold with nonzero curvature. This example has been generalized to almost all noncompact symmetric spaces of rank one and to a countable family of distinct cost functionals [26-27]. The example of stochastic control in a sphere is given in [28]. It can be generalized to other compact symmetric spaces of rank one. Even in Euclidean space there are relatively few explicitly solvable stochastic control problems (e.g. [3, 4, 32, 38, 53]).

§8. The conditional probability density function for the estimation problem here was given by the author in [23]. Various generalizations of it exist. An example in $SU(2)$ is given in [23] and an example for an arbitrary compact semisimple Lie group is given in [25]. Other examples can be solved for real semisimple Lie groups that are applicable to the study of families of transfer functions in system theory.

REFERENCES

[1] ATIYAH, M., BOTT, R. and PATODI, V. K., (1975). *On the heat equation and the index theorem*, Invent. Math. 14, 279-330.

[2] BABBITT, D. G., (1970). *Wiener integral representations for certain semigroups which have infinitesimal generators with matrix coefficients*, J. Math. Mech. 19, 1051-1067.

[3] BENES, V. E., SHEPP, L. A. and WITSENHAUSEN, H. S., (1980). *Some solvable stochastic control problems*, Stochastics 4, 39-83.

[4] BENSOUSSAN, A. and VAN SCHUPPEN, J. H., (1985). *Optimal control of partially observable stochastic systems with an exponential-of-integral performance index*, SIAM J. Control Optim. 23, 599-613.

[5] BISMUT, J. M., (1984). *The Atiyah-Singer theorems: a probabilistic approach I*, J. Funct. Anal. 57, 54-99.

[6] CAMERON, R. H. and MARTIN, W. T. (1944). *Transformation of Wiener integrals under translations*, Ann. Math. 45, 386-396.

[7] CARTAN, É., (1928, 2nd ed. 1946). *Leçons sur la géométrie des espaces de Riemann*, Gauthier-Villars, Paris.

[8] DAVIS, M. H. A. and SPATHOPOULOS, M. P. *A stochastic maximum principle for diffusions on manifolds*, preprint.

[9] DEBIRD, A., GAVEAU, B. and MAZET, E., (1976). *Théorèmes de comparaison en géométric riemannienne*, Publ. RIMS, Kyoto Univ. 12, 391-425.

[10] DUNCAN, T. E., (1975). *Some stochastic systems on manifolds*, Lecture Notes in Econ. and Math. Systems, 107, 262-270, Springer, Berlin-New York.

[11] DUNCAN, T. E., (1975). *A note on some laws of the iterated logarithm*, J. Multivariate Anal. 5, 425-433.

[12] DUNCAN, T. E., (1979). *Stochastic systems in Riemannian manifolds*, J. Optimization Theory Appl. 27, 399-426.

[13] DUNCAN, T. E., (1976). *Stochastic integrals in Riemann manifolds*, J. Multivariate Anal. 6, 397-413.

[14] DUNCAN, T. E., (1977). *Dynamic programming optimality criteria for stochastic systems in Riemannian manifolds*, Appl. Math. Optim. 3, 191-208.

[15] DUNCAN, T. E., (1977). *Some filtering results in Riemann manifolds*, Information and Control 35, 182-195.

[16] DUNCAN, T. E., (1978). *Optimal control of stochastic systems in a sphere bundle*, Lecture Notes in Math. 695, 51-62, Springer, Berlin-New York.

[17] DUNCAN, T. E., (1978). *Estimation for jump processes in the tangent bundle of a Riemann manifold*, Appl. Math. Optim. 4, 265-274.

[18] DUNCAN, T. E., (1979). *The heat equation, the Kac formula and some index theorems*, Lecture Notes in Pure and Appl. Math 48, 57-76, Marcel Dekper, New York - Basel.

[19] DUNCAN, T. E., (1980). *Optimal control of continuous and discontinuous processes in a Riemannian tangent bundle*, Lecture Notes in Math. 794, 396-411, Springer, Berlin-New York.

[20] DUNCAN, T. E., (1981). *Stochastic filtering in manifolds*, Proc. Intern. Fed. Auto. Control World Congress, Pergamon.

[21] DUNCAN, T. E., (1983). *Some geometric methods for integration in manifolds*, Proc. APSM Workshop on System Geometry, System Indentification and Parameter Estimation, 63-72, Math Sci. Press, Brookline.

[22] DUNCAN, T. E. *Brownian motion and affine Lie algebras*, accepted for publication in J. Functional Analysis.

[23] DUNCAN, T. E., (1987). *Explicit solutions for an estimation problem in manifolds associated with Lie groups.* Differential Geometry: The Interface between Pure and Applied Mathematics (M. Luksic, C. Martin and W. Shadwick, eds.), Contemporary Mathematics 68, 99-110, Amer. Math. Soc., Providence.

[24] DUNCAN, T. E., (1987). *A solvable stochastic control problem in hyperbolic three space*, Systems and Control Letters 8, 435-439.

[25] DUNCAN, T. E., (1988). *Some estimation problems in compact Lie groups*, to appear in Systems and Control Letters 10.

[26] DUNCAN, T. E., (1987). *Some solvable stochastic control problems in real hyperbolic spaces*, to appear in Proc. 1987 Conf. on Math. Theory of Networks and Systems, North Holland.

[27] DUNCAN, T. E. *Some solvable stochastic control problems in noncompact, rank one symmetric spaces*, preprint.

[28] DUNCAN, T. E. *A solvable stochastic control problem in spheres*, to appear in Geometry of Random Motion (R. Durrett and M. Pinsky, eds.) Contemporary Mathematics, Amer. Math. Soc., Providence.

[29] DYNKIN, E. B., (1968). *Diffusion of tensors*, Dokl. Akad. Nauk SSSR 179 (1968), 1264-1267. English translation Soviet Math 9, 532-535.

[30] EELLS, J. and ELWORTHY, K. D., (1971). *Wiener integration on certain manifolds*, Problems in Non-Linear Analysis, ed. G. Prodi, 67-94, Centro Intern. Math. Estivo, IV Ciclo., Rome.

[31] ELWORTHY, K. D., (1982). *Stochastic Differential Equations on Manifolds*, London Math. Soc. Lecture Note Ser. 70, Cambridge Univ. Press, Cambridge.

[32] FLEMING, W. H. and RISHEL, R. W., (1975). *Deterministic and Stochastic Optimal Control*, Springer, Berlin - New York,.

[33] FRENKEL, I. B., (1984). *Orbital theory for affine Lie algebras*, Invent. Math. 77 (1984), 301-352.

[34] GANGOLLI, R., (1964). *On the construction of certain diffusions on a differentiable manifold*, Z. Wahr. verw. Geb. 2, 406-419.

[35] GETZLER, E., (1986). *A short proof of the Atiyah-Singer Index Theorem*, Topology 25 , 111-117.

[36] GILKEY, P. B., (1973). *Curvature and the eigenvalues of the Laplacian for elliptic complexes*, Advances in Math. 10, 344-382.

[37] GIRSANOV, I. V., (1960). *On transforming a certain class of stochastic processes by absolutely continuous substitution of measures*, Theor. Probab. Appl. 5, 285-301.

[38] HAUSSMANN, U. G., (1981). *Some examples of optimal stochastic controls or: the stochastic maximum principle at work*, SIAM Rev. 23, 292-307.

[39] HELGASON, G., (1978). *Differential Geometry, Lie Group, and Symmetric Spaces*, Academic Press, New York.

[40] IKEDA, N. and MANABE, S., (1979). *Integral of differential forms along the path of diffusion processes*, Publ. RIMS, Kyoto Univ. 15, 827-852.

[41] ITÔ, K., (1950). *Stochastic differential equations in a differentiable manifold*, Nagoya Math. J. 1, 35-47.

[42] ITÔ, K., (1950). *Stochastic differential equations in a differentiable manifold (2)*, Mem. Coll. Sci. Univ. Kyoto Math. 28, 81-85.

[43] ITÔ, K., (1963). *The Brownian motion and tensor fields on Riemannian manifolds*, Proc. Intem. Congr. Math. Stockholm, 536-539.

[44] ITÔ, K., (1975). *Stochastic parallel displacement, Probabilistic Methods in Differential Equations*, Lecture Notes in Math. 451, 1-7, Springer, Berlin – New York.

[45] KAC, M. (1951). *On some connections between probability theory and differential and integral equations*, Proc. Second Berkeley Symp. math. Statist. Prob., 189-215, Univ. California press, Berkeley.

[46] KAC, V. G., (1968). *Simple irreducible graded Lie algebras of finite growth*, Math, USSR-Izv. 2, 1271-1311.

[47] KOBAYASHI, S. and NOMIZU, K. (1963). *Foundations of Differential Geometry*, Vol. 1, John Wiley - Interscience, New York.

[48] KOSTANT, B. (1976). *On Macdonald's η-function, the Laplacian and generalized exponents*, Advances in Math. 20, 179-212.

[49] MALLIAVIN, P. (1974). *Formule de la moyenne, calcul de perturbations et théorèmes d'annulation pour les formes harmonique*, J. Funct. Analy. 17 , 274-291.

[50] MALLIAVIN, P. (1978). *Géométrie différentielle stochastique*, Les Presses de l'Université de Montréal, Montréal.

[51] MACDONALD, I. G., (1972). *Affine root systems and Dedekind's η-function*, Invent. Math. 15, 91-143.

[52] MCKEAN, H. P. and SINGER, I. M., (1967). *Curvature and the eigenvalues of the Laplacian*, J. Diff. Geom. 1, 43-69.

[53] MERTON, R. C., (1971). *Optimum consumption and portfolio rules in a continuous-time model*, J. Economic Theory 3, 373-413.

[54] MEYER, P. A., (1981). *A differential geometric formalism for the Itô calculus*, Lecture Notes in Math. 851, 256-270, Springer, Berlin – New York.

[55] MOODY, R. V., (1968). *A new class of Lie algebras*, J. Algebra 10, 211-230.

[56] NG, S. K. and . CAINES, P. E., (1985). *Nonlinear filtering in Riemannian manifolds*, IMA J. Math. Control Inform. 2, 25-36.

[57] NOMIZU, K., (1956). *Lie Groups and Differential Geometry*, Publ. Math. Soc. Japan, Tokyo.

[58] PATODI, V. K., (1971). *Curvature and the eigenforms of the Laplace operator*, J. Diff. Geom. 5, 233-249.

[59] PINKSY, M., (1976). *Isotropic transport process on a Riemannian manifold*, Trans. Amer. Math. Soc. 218, 353-360.

[60] PINKSY, M., (1978). *Stochastic Riemannian geometry*, Probabilistic Analysis and Related Topics (ed. by A.T. Bharucka-Reid) Academic Press, New York.

[61] PONTIER, M. and SZPIRGLAS, J., (1985). *Filtrage non- lineaire avec observation sur une variété*, Stochastics 15, 121-148.

[62] STROOCK, D. W., (1970). *On certain systems of parabolic equations*, Communications Pure Appl. Math. 23, 447-457.

[63] YAU, S. T., (1976). *Some function-theoretic properties of complete Riemannian manifolds and their applications to geometry*, Indiana Univ. Math. J. 25, 659-670.

[64] YAU, S. T., (1978). *On the heat kernel of a complete Riemannian manifold*, J. Math. Pures Appl. 57, 191-201.

[65] YANO, K. and BOCHNER, S., (1953). *Curvature and Betti Numbers*, Ann. Math. Studies 32, Princeton Univ. Press, Princeton.

5

CONTROLLED MARKOV CHAINS ON A COUNTABLE STATE SPACE: SOME RECENT RESULTS

Vivek S. Borkar

5.1 INTRODUCTION

This is a brief survey of some recent results on controlled Markov chains, the details of which appear in [1]-[3]. The main idea in these works is to pose the control problem as an optimization problem on an appropriate set of probability measures. For cost criteria such as discounted cost, finite time horizon cost and cost up to an exit time, this approach allows us to recover known results from a different vantage point and appears promising for certain constrained optimization problems [1]. For the long-run average cost, the approach leads to very general results that seem to subsume most known results on this problem [2], [3]. More on this later.

In the next section, the notation of [1]-[3] is described. Section 5.3 reviews the results of [1] concerning discounted cost, finite horizon and exit time problems. Sections IV and V are based on [2], [3] and describe the long-run average cost control problem. Section IV deals with the existence of optimal stable stationary strategies and Section V the corresponding dynamic programming equations. Section 5.6 concludes with a brief list of futher possibilities.

The material presented here is based on the work I did while visiting the Institute for Mathematics and its Applications, University of Minnesota, and Systems Research Center, University of Maryland at College Park, where it was supported by NSF grant CDR-85-00108. It formed the core of a set of lectures I gave at the latter institution.

I would like to thank Professor Prakash Narayan for inviting me to give these lectures and Professor John Baras for inviting me to submit this article for the proceedings.

5.2 PRELIMINARIES

Let $x_n, n = 1, 2, \ldots,$ be a controlled Markov chain on a countable state space $S = \{1, 2, \ldots\}$. Let $D(i), i \varepsilon S$, be a prescribed collection of compact metric spaces which shall serve us as the 'control sets.' The dynamics of $\{X_n\}$ will be dictated (in a manner soon to be specified) by a transition matrix $P_u = [[p(i, j, u_i)]], i, j \varepsilon S$, indexed by the control vector $u = [u_1, u_2, \ldots] \varepsilon \Pi\, D(i)$. The functions $p(i, j, .)$ are assumed to be continuous. By replacing each $D(i)$ by Π D(j) and $p(i, j, .)$ by its composition with the projection ii D(j) $\rightarrow$ D(i), one may assume that all $D(i)$'s are replicas of the same compact metric space D. We do so and then define L as the countable product of copies of D with the product topology.

For any Polish space $Y, M(Y)$ will denote the space of probability measures on Y with the topology of weak convergence. For $n = 1, 2, \ldots, \infty, Y^n$ will denote the n-times product of Y with itself.

A control strategy (CS) is a sequence $\{\xi_n\}$, $\xi_n = [\xi_n(1), \xi_n(2), \ldots]$ of L-valued random variables such that for $i \varepsilon S$, $n \geq 1$,

$$P(X_{n+1} = i / X_m, \xi_m, m \leq n) = p(X_n, i\, \xi_n(X_n)). \tag{5.2.1}$$

The controlled Markov chain $\{X_n\}$ is said to be governed by the $CS\, \{\xi_n\}$ whenever (5.1.1) holds. If for each n, ξ_n is independent of X_m, $m \leq n$, and ξ_m, $m < n$, we call $\{\xi_n\}$ a Markov randomized strategy (MRS). If in addition ξ_n, $n \geq 1$, are identically distributed, call it a stationary randomized strategy (SRS). An MRS (SRS) for which the law of each ξ_n is a Dirac measure will be called a Markov strategy (MS) (stationary strategy (SS) resp.). The motivation for this nomenclature is self-evident.

In case of the long-run average cost problem, we are also interested in the stability-based classification of some of the control strategies introduced above. Thus we call an SRS a stable stationary randomized strategy (SSRS) if the resultant Markov chain is positive recurrent. A stable stationary strategy (SSS) is defined analogously. We assume throughout that the chaim has a single communicating class under any

SRS. Thus, the recurrence properties are independent of the initial condition and under any SSS or SSRS, the chain has a unique invariant probability measure.

If the common law of $\xi_n, n \geq 1$, for an SRS $\{\xi_n\}$ is $\phi \; \varepsilon \; M(L)$, we denote the SRS by $\gamma[\phi]$. We shall be interested only in the law of the $S \times D$ - valued process $\{(X_n, \xi_n(X_n))\}$, $n \geq 1$. Given this, it is not difficult to see that for an SRS $\gamma[\phi]$, ϕ may be assumed to be of a product form. This is, $\phi, ., \varepsilon \, M(L)$ is given by $\phi = \overset{\text{II}}{_i} \phi_i$, where $\phi_i \; \varepsilon \; M(D)$. Under $\gamma[\phi]$, $\{X_n\}$ will be a Markov chain with stationary transition matrix $P[\phi] = [[\int p(i,j,u) \, \phi_i \, (du)]]$. If $\gamma[\phi]$ is an SS with $\phi =$ the Dirac measure at $\xi \, \varepsilon \, L$, denote it by $\gamma\{\xi\}$ and the corresponding transition matrix by $P\{\xi\} = P_\xi$.

Let $h : S \to R^+$, $k : S \times D \to R^+$, $\ell = IN \times S \times D \to R^+$ be continuous functions. The various cost criteria one typically seeks to minimize over all CS are the following:

$$E\left[\sum_{n=1}^{\infty} \beta^n K(X_n, \xi_n(X_n))\right], \tag{C1}$$

for some $\beta \varepsilon (0, 1)$. This is the discounted cost control problem.

$$E\left[\sum_{n-1}^{N-1} \ell(n, X_n, \xi_n(X_n)) + h(X_N)\right], \tag{C2}$$

for some $N, 1 \leq N \infty$. This is the finite horizon control problem.

$$E\left[\sum_{n=1}^{\tau-1} k(X - N, \xi_n(X_n)) + h(X_\tau)\right] \tag{C3}$$

where for some prescribed finte subset A of S,

$$\tau = \min\{n \geq 1/X_n \not\varepsilon A\}.$$

(By convention, $\tau = \infty$ if $X_n \, \varepsilon \, A$ for all n.). This is called the 'control up to a first exit time.'

$$\lim_{n \to} \inf_{\infty} \frac{1}{n} \sum_{m=1}^{n} k(X_m, \xi_m(X_m)) \tag{C4}$$

This is the 'long-run average cost control problem.'

A CS will be said to be optimal if it minimizes the appropriate cost. (In case of (C4), replace 'minimizes' by 'a.s. minimizes').

5.3. THE COSTS (C1) - (C3)

The results of this section evolve from the following basic fact, proved in [1]. For a fixed initial law for X_1, the set of attainable laws of $\{(X_1, \xi(X_1)), X_2, \xi_2(X_2)), \ldots\}$, viewed as a subset $M((S \times D)\infty)$, is convex and compact (resp. compact) as the CS varies all CS (resp. all MRS, MS, SRS, or SS).

To each of (C1) - (C3), we assign an appropriate 'occupation measure' as follows: For (C1), define the 'discounted occupation measure' to be $\nu \varepsilon\ M(S \times D)$ defined by

$$\int f d\nu = E\left\{\sum_{n=1}^{infty} \beta^n f(X_n,\ \xi_n(Xn))\right\} \text{ for } f \varepsilon C_b\ (S \times D).$$

For (C2), the 'finite horizon occupation measure' is $\nu \varepsilon\, M(\{1, 2, \ldots N\} \times S \times D)$ defined by

$$\int f d\nu = E\left[\sum_{n=1}^{N} f(n, Xn, \xi_n(X_n))\right] /N \text{for } f \varepsilon C_b\,(\{1,\ 2,\ \ldots,\ N\} \times S \times D).$$

For (C3), the 'occuption measure up to the first exist from A' is $\nu \varepsilon\ M(A \times D)$.

Let $B_1(F), B_2(F), B_3(F)$ denote respectively the sets of attainable ν's in the above three cases as the CS varies over a class F of CS. The main result of [1] is the following.

Theorem 5.3.1

a. B_1 (CS) = B_1 (SRS) and is compact convex in $M(S \times D)$ with its extreme points lying in B_1 (SS) which itself is compact.

b. B_2(CS) = B_2 (MRS) and is compact convex in $M(\{1,\ 2., \ldots, N\} \times S \times D)$ with its extreme points lying in B_2 (MS) which itself is compact.

c. B_3(CS) = B_3 (SRS) and is compact convex in $M(A \times D)$ with its extreme points lying in B_3 (SS) which itself is compact.

Note that in each of (C1) - (C3), the cost can be expressed as an integral of some function with repsect to the corresponding occupation measure. The above theorem then allows us to deduce the following.

Corollary 5.3.1: There exists an optimal SS for (C1) and (C3) and an optimal MS for (C2).

Thus we recover a result which was hitherto obtained via dynamic programming equations, by purely convex analytic means. This alternative line of attack seems promising (at least as a starting point) for more complex situations such as those arising from constrained optimization problems. This aspect is still at a conceptual stage and a brief speculative discussion appears in [1].

5.4 THE COST (C4): THE EXISTENCE RESULTS

The structure of the control problem corresponding to (C4) differs drastically from those considered above because it is the 'ergodic' behavior of the process and not its finite time behavior that is at stake. The key results on which the existence theory is based are Theorems 5.4.1 and 5.4.2 below. We start with some preliminaries.

Let $\bar{S} = SU\{\infty\}$ denote the one point compactification of S. Each $\nu \varepsilon M(\bar{S} \times D)$ has a decomposition: $\nu(A) = \delta_\nu \nu_1(A \cap (S \times D)) + (1 - \delta_\nu)\nu_2(A \cap (\infty \times D))$, a Borel in $\bar{S} \times D$, where $\delta_\nu \; \varepsilon \; [0,1]$, $\nu_1 \; \varepsilon \; M(S \times D)$ and $\nu_2 \varepsilon M(\{\infty\} \times D)$. Furthermore, $\nu - 1$ is uniquely specified when $\delta_\nu > 0$. Disintegrate ν_1 as

$$\nu_1(\{i\},\, du) = m_\nu(i)\phi_{\nu i}(du), \quad i \varepsilon S,$$

where $m_\nu \varepsilon M(S)$ is the image of ν_1 under the projection $S \times D \to S$ and $i \varepsilon S \to \phi_{\nu i} \varepsilon M(D)$ is the regular conditional law. (The latter is specified M_ν a.s. Pick any one representative of this a.s. equivalence class.) Let$\phi_\nu = \prod_i \phi_{\nu i}$.

Let $\pi[\phi]$, $\pi\{\xi\}$ denote respectively the unique invariant probability measures under an SSRS $\gamma[\phi]$ or an SSS $\gamma\{\xi\}$. For $n \geq 1$, define the $M(\bar{S} \times D)$ - valued processes $\{\nu_n\}$ by

$$\int f d\nu_n = \frac{1}{n} \sum_{m=1}^{n} f(X_m,\, \xi_m\,(X_m)),\; f \varepsilon\; C(\bar{S} \times D),$$

where $\{X_n\}$ is a chain governed by $\{\xi_n\}$.

Theorem 5.4.1

$\nu_n \to \{\nu \varepsilon M(\bar{S} \times D) | \gamma_\nu = 0\} U \{\nu \varepsilon M(\bar{S} \times D) | m_\nu = \pi[\phi_\nu]\}$ as in $M(\bar{S} \times D)$.

For an SSRS $\gamma[\phi]$, define $\hat{\pi}[\phi] \,\varepsilon\, M(S \times D)$ as: $\hat{\pi}\,[\phi]\,(\{i\},\,du) = \pi\,[\phi](i)\,\phi_i(du)$, $\phi_i \varepsilon M(D)$ being the image of ϕ under the projection from L into its i-th factor space. For an SSS $\gamma\{\xi\}$, define $\hat{\pi}\{\xi\}\varepsilon\, M(S \times D)$ analogously.

Theorem 5.4.2

The sets $\{\hat{\pi}[\phi]\,\gamma\,[\phi]$ an SSRS$\}$ and $\{\hat{\pi}\{xi\}|\gamma\{\xi\}\text{SSS}\}$ are closed in $M(S \times D)$ and the former set is convex with its extreme points in the latter set.

These two results form the key steps in the proofs of the existence theorems below.

Theorem 5.4.3

Suppose there exists an SSS $\gamma\{\xi\}$ such that $\int k d\hat{\pi}\{\xi\} < \infty$ and furthermore, k satisfies:

$$\lim_{i \to} \inf_{\infty} \inf_u k(i,u) > \inf_{\text{all SSRS}} \int k d\hat{\pi}[\phi].$$

Then an optimal SSS exists.

Theorem 5.4.4

Suppose there exists an SSS $\gamma\{\xi\}$ such that $\int k d\hat{\pi}\{\xi\} <$ and furthermore, k satisfies:

$$\sup E[\tau^2/X_1 + 1] < \infty (5.4.1)$$

where τ is the first return time to state 1 and the supremum is over all CS. Then all SS are SSS, $\{\hat{\pi}\{\xi\}\,|\xi \varepsilon L\}$ is compact in $M(S \times D)$ and an optimal SSS exists.

Remarks.

(1) Sufficient conditions for (5.4.1) to hold are given in [2]. These are either conditions on the graph of the chain or require the existence of a suitable 'Liapunov function.'

(2) If the supremum in (5.4.1) is over all SS, one is still assured of an optimal SSS within the class of all SRS.

5.5 THE COST (C4): THE DYNAMIC PROGRAMMING EQUATIONS

The results of this section are from [3]. Continuing with the set-up of the preceding section, we assume that an SSS $\gamma\{\xi_0\}$ exists. Given an SSS $\gamma\{\xi\}$, define an infinite vector, $V\{\xi\} = [V\{\xi\})(1), V\{\xi\}(2), \ldots]^T$ by

$$V\{\xi\}(i) = E\left[\sum_{n=1}^{\tau-1} (k(X_n, \xi_n(X_n)) - \alpha) / X_1 = i\right]$$

where

$$\tau = \min\{n \geq 1 | X_n = 1\}$$
$$\alpha = \underset{\text{all SSS}}{\inf} \int kd\hat{\pi}\{\xi\} = \int kd\hat{\pi}\{xi_0\}$$

and the expectation is with respect to the law under $\gamma\{\xi\}$. Let $1_c = [1, 1, \ldots]^T$, $U = [[\delta_{ij}]], i, j, = 1, 2, \ldots$, (i.e., U = the infinite identity matrix), and $Q_\xi = [k(1, \xi(1)), k(2, \xi(2)), \ldots]^T$.

Throughout this section we impose the following additional hypothesis, called the 'stability under local perturbation': If $\gamma\{\xi\}$ is an SSS, so is $\gamma\{\xi'\}$ for any $\xi' \varepsilon L$ such that $\xi'(i) \neq \xi(i)$ for at most finitely many $i \varepsilon S$.

Theorem 5.5.1 There exists an infinite column vector $V = [V(1), V(2), \ldots]^T$ such that

$$\alpha 1_c = \underset{\xi}{\min} ((P\{\xi\} - U)V + Q_x i$$

where the minimum is termwise. Futhermore, an SS $\gamma\{\xi\}$ is optimal if and only if ξ attains the minimum in (5.5.1).

Remarks.

(1) (5.5.1) are called the dynamic programming equations. Note that V could be changed by a constant multiple of 1_c without altering the validity of (5.5.1).

(2) An explicit choice of V is $V\{\xi_0\} + K1_c$ for any constant K. This vector is unaltered if $\gamma\{\xi_0\}$ is replaced by any other optimal SSS and changed only by a constant multiple of 1_c if our choice of state '1' in the definition of τ is changed.

(3) Under the hypotheses of Theorem 4.4, all SS are SSS and hence in the last claim of the theorem above, SSS may be repleced by SS. The same holds true under

the hypotheses of Theorem 5.4.3, though not in such an obvious manner: One can prove that if an SS $\gamma\{\xi\}$ is either optimal of is such that ξ attains the minimum in (5.5.1), it must be stable.

Now define an SSS $\gamma\{\xi\}$ to be locally optimal if for any $\xi' \varepsilon L$ such that $\xi'(i) \neq \xi(i)$ for at most finitely many $i \varepsilon S$,

$$\int k d\hat{\pi}\{\xi'\} \geq \int k d\hat{\pi}\{\xi\}.$$

Theorem 5.5.2 Suppose an SSS $\gamma\{\xi\}$ satisfies

$$\left(\int k d\hat{\pi}\{\xi\}\right)1_c = \min_{\psi}\left((P\{\psi\} - U)V\{\psi\} + Q\psi\right).$$

Then $\gamma\{\xi\}$ is locally optimal.

Remarks.

(1) The minimum in (5.5.2) will be attained at ξ.

(2) The converse implication also holds.

(3) (5.5.2) does not involve the 'unknowns' α and V as in (5.5.1). However, it pertains to only local optimality as opposed to optimality. In some cases, e.g. under the hypotheses of Theorem 5.4.4 with bounded k, local optimality can be shown to be equivalent to optimality.

So far we have not touched the issue of characterizing the pair (α, V) from all the pairs (c, W), $c \varepsilon R, W = [W(1), W(2), \ldots]^T$, satisfying

$$c1_c = \min_{\xi}((P\{\xi\} - U)W + X_x i). \tag{5.5.3}$$

A partial result is as follows:

Theorem 5.5.3 Under the hypotheses of Theorem 5.4.3, $(\beta, V\{\xi_0\})$, is the unique element of $\{(c, W) | (c, W)$ satisfies (5.5.3), $W(1) = 0$, $\inf W(i) > -\infty\}$ that corresponds to the least value of c.

5.6 CONCLUSIONS

In conclusion, we list a few problems suggested by the foregoing:

(1) Is (5.4.1) always true when all SS are SSS?

(2) Does local optimality always imply optimality in the set-up of Section V?

(3) Find a suitable characterization of $(\beta, V\{\xi_0\})$ along the line of Theorem 5.5.3 (some modifications being obviously essential) under the hypotheses of Theorem 5.4.4.

(4) Can the hypothesis of stability under local perturbation be relaxed for the results of Section V?

REFERENCES

[1] BORKAR, V. S., *A fresh look at Markov decision processes*, preprint, Systems Research Center, University of Maryland, College Park, 1987.

[2] BORKAR, V. S., *Control of Markov chains with long-run average cost criterion*, to appear in Proceedings of the I.M.A. Workshop on Stoch. Diff. Systems with Appl. in Electrical/Computer Engg., Control Theory and Op. Research (W. Fleming and P. L. Lions, Eds.,), Springer, 1987.

[3] BORKAR, V. S., *Control of Markov chains with long-run average cost criterion* II, preprint, Systems Research Center, University of Maryland, College Park, 1987.

6

MARTINGALE PROBLEMS FOR CONSTRAINED MARKOV PROBLEMS

Thomas G. Kurtz

Let E be a complete, separable metric space, and let $A \subset B(E) \times B(E)$. Then a measurable stochastic process X is a solution of the martingale problem for A if there exists a filtration $\{\mathcal{F}_t\}$ such that X is adapted to $\{\mathcal{F}_t\}$ and

$$f(X(t)) - \int_0^t g(X(s))ds \tag{6.0.1}$$

is an $\{\mathcal{F}\}_t$ - martingale for all $(f,g) \in A$. Ordinarily, A will be the graph of a single-valued operator and we will write Af in place of g. We are interested in characterizing processes X which behave as a solution of the martingale problem for A as long as the process remains in an open subset $E_0 \subset E$ and is constrained to remain in $\bar{E}_0$, the closure of E_0, by modifying the behavior of the process on E_0^c.

For diffusion processes (that is, the case in which A is a second order elliptic operator) this problem has been extensively studied using a variety of approaches. Ventsel' (1959) considered the problem of characterizing the positive semigroups on $C(\bar{E}_0)$ whose infinitesimal generators are closures of restrictions of a given elliptic operator. Under compactness and smoothness conditions on the domain, he showed that any C^2 function in the domain of the infinitesimal generator satisfied boundary conditions of the form $Bf(x) = 0, x \in \partial E_0$, where B is a second order integral differential operator. This analytic approach has been pursued further by Sato and Ueno (1965), Bony, Courrége, and Priouret (1968), and Taira (1977). Skorohod (1961) and Ikeda (1961) formulated stochastic differential equations for diffusions in bounded domains and this approach has been followed by Watanabe (1971), Tanaka (1979), and Lions and Sznitman (1984). Stroock and Varadhan (1971) and Anderson (1976) consider a submartingale problem in which (6.01) is required to be a submartingale for every (f, g) for which $Bf \geq 0$ on ∂E_0.

In these latter two papers the authors show that for any solution of their submartingale problem there exists a nondecreasing process λ which increases only when $X \in \partial E_0$ such that

$$f(X(t)) - \int_0^t Af(X(s))ds - \int_0^t Bf(X(s))d\lambda(S) \tag{6.0.2}$$

is a martingale. This "martingale problem"is explicitly formulated as such in Mikulyavichyus (1980 a, b, 1982) for A and B integro-differential operators on C^2 manifolds and in Ikeda and Watanabe (1981). It is this martingale problem we wish to consider in a more general context. Note, for a stochastic process X and a set Γ, we will say that a nonnegative process λ with $\lambda(0) = 0$ is a local time for X in Γ, if λ is nondecreasing and λ increases only when X is in Γ, i.e.,

$$\lambda(t) = \int_0^t \chi_{\{X(s) \in \Gamma\}} d\lambda(s). \tag{6.0.3}$$

In the process of attempting to construct solutions of the martingale problem given by (6.0.2), it will be seen that we really need to consider a more general formulation of the problem. We are led to this more general problem by our method of constructing solutions; however, once discovered, we will see that this formulation of the problem is in fact useful, i.e., not every constrained Markov process is a solution of a martingale problem of the form in (6.0.2). The method of construction involves the introduction of a "patchwork martingale problem"which is discussed in Section 1.

To simplify the formulation of some of the relative compactness results we are interested in, we will assume that the original state space E is compact. This is not particularly restrictive since if E is locally compact and $A \subset \hat{C}(E) \times \hat{C}(E)$ (for example, if A is the generator of a Feller process), then we can replace E by its one-point compactification $E^\triangle$, and extend the functions in A continuously to $E^\triangle$ so that any solution of the original martingale problem gives a solution of the new problem. (See Ethier and Kurtz (1986) Section 4.3 for conditions under which the converse will hold.)

6.1 THE PATCHWORK MARTINGALE PROBLEM

We begin with a lemma which serves to introduce the patchwork martingale problem.

6.1.1 Lemma Let E be compact, and let $B_0, B_1, \ldots, B_m \subset C(E) \times C(E)$ be dissipative operators, each containing $(1,0)$ and having a common domain D which is dense in $C(E)$. Let $E_0, \ldots, E_m$ be disjoint Borel subsets of E satisfying $E = \cup_{i=0}^m E_i$. Then for each $\nu \in \wp(E)$ there exists a process $(X, \lambda_0, \ldots, \lambda_m)$ such that X has sample paths in $D_E[0,\infty)$ and initial distribution ν, for each $i = 0, \ldots, m$, λ_i is nondecreasing and

$$\int_0^t \chi_{\{X(s)\in \bar{E}_i\}} d\lambda_i(s) = \lambda_i(t), \text{ a.s.,} \tag{6.1.1}$$

$\sum_{i=0}^m \lambda_i(t) = t$, a.s., and there exists a filtration $\mathcal{F}_t$ such that

$$f(X(t)) - \sum_{i=0}^m \int_0^t B_i f(X(s)) d\lambda_i(s) \tag{6.1.2}$$

is an $\{\mathcal{F}_t\}$-martingale for each $f \in D$.

6.1.2 Remark The process X (or when convenient, $(X, \lambda_0, \ldots, \lambda_m)$) will be called a solution of the patchwork martingale problem for $(B_0, E_0, \ldots, B_m, E_m)$.

Proof: For each $i = 0, \ldots, m$ there exists a sequence of one-step transition functions $\{\mu_i^n\}$ on E such that $B_i^n f(x) \equiv n \int (f(y) - f(x)) \mu_i^n(x, dy)$ satisfies $\lim_{n\to\infty} B_i^n f = B_i F$ (uniformly on E) for every $f \in \mathcal{D}(B_i)$. (See Ethier and Kurtz (1986) Lemma 4.5.3.) For $n = 1, 2, \ldots$, define μ^n by

$$\mu^n(x, \Gamma) = \sum_{i=0}^m \chi_{E_i}(x) \mu_i^n(x, \Gamma),$$

and let $\{Y_n(k) : k \geq 0\}$ be a Markov chain with initial distribution ν and transition function μ^n. Define $X_n(t) = Y_n([nt])$ and

$$\lambda_i^n(t) = \int_0^t \chi_{E_i}(X_n(s)) ds.$$

Then

$$
\begin{aligned}
f(X_n(t)) - \sum_{k=0}^{[nt]-1} \int (f(y) - f(Y_n(k)))\mu^n(Y_n(k), dy) \\
= f(X_n(t)) - \int_0^{[nt]/n} \sum_{i=0}^{m} \chi_{E_i}(X_n(s))B_i^n f(X_n(s))ds \qquad (6.1.3) \\
= f(X_n(t)) - \sum_{i=0}^{m} \int_0^{[nt]/n} B_i^n f(X_n(s))d\lambda_i^n(s)
\end{aligned}
$$

is an $\{\mathcal{F}_t^{X_n}\}$-martingale for all $f \in B(E)$. Since D is dense in $C(E)$ and $||B_i^n f||$ is bounded for $f \in D$, Theorems 3.9.1 and 3.9.4 of Ethier and Kurtz (1986) give the relative compactness of $\{X_n\}$ in $D_E[0,\infty)$ and the Lipschitz continuity of the λ_i^n gives the relative compactness of $\{\lambda_i^n\}$ in $C_{\Re}[0,\infty)$. Consequently, there exists a subsequence of $\{(X_n, \lambda_0^n, \dots, \lambda_m^n)\}$ which converges in distribution to a process $(X, \lambda_0, \dots, \lambda_m)$ in $D_E[0,\infty) \times C_{\Re_m}[0,\infty)$. Note that $\nu_i^n(\Gamma) = \int_0^t \chi_\Gamma(X_n(s))\, d\lambda_i^n(s)$ defines a measure concentrated on E_i, and it follows that $\nu_i(\Gamma) = \int_0^t \chi_\Gamma(X(s))\, d\lambda_i(s)$ will be concentrated on $\bar{E}_i$ which gives (6.1.1).

Next note that

$$
f(X_n(t)) - \sum_{i=0}^{m} \int_0^t B_i f(X_n(s))d\lambda_i^n(s)
$$

converges in distribution to (6.1.2) and, by the definition of B_i^n, is asymptotically equal (in L^1) to (6.1.3). In particular, (6.1.3) converges in distribution to (6.1.2), and it follows that (6.1.2) is an $\{\mathcal{F}_t\}$-martingale where $\mathcal{F}_t = \sigma((X(s), \lambda_0(s), \dots, \lambda_m(s)) : s \le t)$.

Our interest in the above lemma is in taking $B_0 = A$ and $B_1, \dots, B_m$ to be generators for processes on different pieces of the boundary of E_0 which constrain X to $\bar{E}_o$. Note that

$$
\begin{aligned}
f(X(t)) - \int_0^t Af(X(s))d\lambda_0(s) - \sum_{i=0}^{m} \int_0^t B_i f(X(s))d\lambda_i(s) \\
= f(X(t)) - \int_0^t Af(X(s))ds - \sum_{i=0}^{m} \int_0^t (B_i - A)f(X(s))d\lambda i(s)
\end{aligned}
\qquad (6.1.4)
$$

so X is a solution of the (ordinary) martingale problem for $A_1 = \{(f,g) \in A : f \in D,$ $B_i f = Af$ on E_i, $i = 1, \dots, m\}$.

6.1.3 Example Let $E = \Re$ (or $[-\infty, \infty]$ to give compactness), $m = 1$, $Af = \frac{1}{2}f''$, $B_f = f'$ for $f \in D = \{f : f, f', f'' \in \bar{C}(-\infty, \infty)\}$. and have limits at $\{.\infty\}$, $E_0 = (0, \infty)$, and $E_1 = (-\infty, 0]$. It is not difficult to see that the process X constructed above is continuous and that $X(t) \in [0, \infty)$ for all $t \geq 0$ a.s. (this is a consequence of the next lemma). Therefore X is a Brownian motion on $[0, \infty)$ with boundary condition $\frac{1}{2}f''(0) = f'(0)$, that is X is a Brownian motion with a sticky boundary.

6.1.4 Lemma Let X be a solution of the patchwork martingale problem for $(A, E_0, B_1, E_1, \ldots, B_m, E_m)$. Suppose there exists $f \in D$ such that $f \geq 0$, $f = Af = 0$ on E_0, $B_i f \leq 0$ on E_i, and there is a closed set $F \supset E_0$ such that $f - B_i f > 0$ on $E_i \cap F^c$. Then $X(0) \in E_0$ implies $X(t) \in F$ for all $t \geq 0$ a.s.

6.1.5 Remark. To see that the lemma applies to Example 6.1.3, take

$$f = \begin{cases} x^4(1 + x^4)^{-1}, & x \leq 0, \\ 0, & x \geq 0, \end{cases}$$

and conclude that $X(0) \in \bar{E}_0$ implies that $X(t) \in \bar{E}_0$ for all $t > 0$ a.s.

Proof: For f satisfying the conditions of the lemma, (6.1.4) is a nonnegative martingale. Consequently, if $X(0) \in E_O$, then $f(X(0)) = 0$ and (6.1.4) is zero for all $t > 0$. (See, for example, Proposition 2.2.15 in Ethier and Kurtz (1986).) But, under the conditions of the lemma, if $X(t) \in F_0$ for some $t > 0$, then the martingale is non-zero for some $t > 0$.

At this point we prove a closely related lemma which will be of interest in Section 6.2.

6.1.6 Lemma Suppose X is a solution of the patchwork martingale problem for $(A, E_0, B_1, E_1, \ldots, B_m, E_m)$ and that $X(t) \in \bar{E}_0$ for all $t \leq 0$ (cf. Lemma 6.1.4). Let $F_1, \ldots, F_m$ be a partition of E with $E_i \subset F_i$ and suppose every solution Z of the patchwork martingale problem for $(B_1, F_1, \ldots, B_m, F_m)$ with $Z(0) \in \partial E_0$ satisfies

$$\inf\{t : Z(t) \in E_0\} = 0 \text{ a.s.} \tag{6.1.5}$$

Then λ_0 is a.s. strictly increasing.

6.1.7 Remark For Example 6.1.3, take $F_1 = (-\infty, \infty)$. Then the patchwork martingale problem for (B_1, F_1) is just the martingale problem for B_1 and the only solution satisfying $Z(0) = 0$ is $Z(t) = t$.

Proof: For $t_0 \geq 0$, let $\eta = \inf\{t > t_0 : \lambda_0(t)\}$. Either λ_0 is a.s. strictly increasing or there exists t_0 such that $P\{\eta > t_0\} > 0$. For such a t_0, define $Z(t) = X(t_0 + t)$ and $\lambda_i^0(t) = \lambda_i(t_0 + t)$ for $t < \gamma \equiv \eta - t_0$. Then

$$f(Z(t \wedge \gamma)) - \sum_{i=1}^{m} \int_0^{t\wedge\gamma} B_i f(Z(s)) d\lambda_i^0(s)$$

is an $\{\mathcal{F}_{t_0+t}\}$-martingale for every $f \in D$ and λ_i^0 increases only when Z is in F_i. It follows that Z can be extended to a solution of the patchwork martingale problem for $(B_1, F_1, \ldots, B_m, F_m)$ (as in the proof of Lemma 4.5.16 of Ethier and Kurtz (1986)). But then $\inf\{t : Z(t) \in E_0\} > \gamma$ which is positive with positive probability contradicting (6.1.5).

6.1.8 Lemma Suppose X is a solution of the patchwork martingale problem for $(A, E_0, B_1, E_1, \ldots, B_m, E_m)$ and that $X(t) \in \bar{E}_0$ for all $t \geq 0$. If there exist $\phi_n \in D$ such that $M \equiv \sup_{n, x\in\bar{E}_0} |\phi_n(x)| < \infty$, and $B_i\phi_n(x) \geq n$, $x \in \bar{E}_i \cap \bar{E}_0$, then λ_0 is strictly increasing.

Proof: Let η be as in the proof of Lemma 6.1.6. Then

$$P\{\eta \geq t_0 + \delta\} \geq \delta^{-1} n^{-1} \sum_{i=1}^{m} E\left[\int_{t_0}^{\eta\wedge(t_0+\delta)} B_i\phi_n(X(s)) d\lambda_i(s)\right] \leq 2\delta^{-1} n^{-1} M,$$

and the lemma follows.

6.1.9 Lemma Let X be a solution of the patchwork martingale problem for $(A, E_0, B_1, E_1, \ldots, B_m, E_m)$. Suppose there exists $f \in D$ and $\epsilon > 0$ such that

$$\sum_{i=1}^{m} \int_0^t B_i f(X(s)) d\lambda_i(s) \geq \epsilon \sum_{i=1}^{m} \lambda_i(t), \qquad t \geq 0. \tag{6.1.6}$$

Then $\lim_{t\to\infty} \lambda_0(t) = \infty$ a.s.

6.1.10 Remark Since Example 6.1.3 satisfies the conditions of Lemma 6.1.4, it also satisfies the conditions of Lemma 6.1.9, i.e., let $f \in D$ satisfy $f'(0) \geq \epsilon$.

Proof: Let $\eta_k = \inf\{t : \lambda_0(t) > K\}$. Then

$$\begin{aligned}
&\frac{2||f|| + K||Af||}{t} \\
&\geq t^{-1}\left\{E\left[f(X(t \wedge \eta_K))\right] - E[f(X(0))] - E\left[\int_0^{t\wedge\eta_k} Af(X(s))d\lambda_0(s)\right]\right\} \\
&\geq t^{-1}E\left[\sum_{i=1}^m \int_0^{t\wedge\eta_K} B_i f(X(s))d\lambda_i(s)\right] \qquad (6.1.7) \\
&\geq t^{-1}\epsilon E\left[\sum_{i=1}^m \lambda_i(t\wedge\eta_k)\right] = t^{-1}\epsilon E\left[t\wedge\eta_K - \lambda_0(t\wedge\eta_K)\right] \\
&\geq \epsilon\frac{t-K}{t}P\{\eta_K > t\}
\end{aligned}$$

and it follows that $\eta_K < \infty$ a.s. for each K, which in turn implies that $\lim_{t\to\infty}\lambda_0(t) = \infty$ a.s.

6.1.11 Remark An examination of the proof of Lemma 6.1.9 shows that it is enough to have a sequence $\{f_n\} \subset D$ such that $\sup_{x,n} f_n(x) < \infty$, $\inf_n E\left[f_n(X(0))\right] > -\infty$,

$\inf_{x,n} Af_n(x) > -\infty$,

$$\sum_{i=1}^m \int_0^t B_i f_n(x(s))d\lambda_i(s) \geq 0, \qquad t \geq 0,$$

and

$$\underline{\lim}_{n\to\infty} \to \sum_{i=1}^m \int_0^t B_i f_n(X(s))d\lambda_i(s) \geq \varepsilon\sum_{i=1}^m \lambda_i(t).$$

By a simple time change, we can generalize the observation made following (6.1.4). For $i = 1, \ldots, m$, let β_i be a nonnegative measurable function on E.

Suppose

$$\gamma(u) \equiv \lambda_0(u) + \sum_{i=1}^m \int_0^u \beta_i(X(s))d\lambda_i(s)$$

is strictly increasing and $\lim_{u\to\infty}\gamma(u) = \infty$. Define $\tau(t) = \tau^{-1}(t)$ and $Y(t) =$

$X(\tau(t))$. Then by the optional sampling theorem

$$\begin{aligned} f(Y(t)) &- \int_0^{\tau(t)} Af(X(s))d\lambda_0(s) - \sum_{i=1}^{m} \int_0^{\tau(t)} B_i f(X(s))d\lambda_i(s) \\ &= f(Y(t)) - \int_0^{\tau(t)} Af(X(s))d\gamma(s) - \sum_{i=1}^{m} \int_0^{\tau(t)} (B_i - \beta_i A)f(X(s))d\lambda_i(s) \\ &= f(Y(t)) - \int_0^{t} Af(Y(s))ds - \sum_{i=1}^{m} \int_0^{t} (B_i - \beta_i A)f(Y(s))d\lambda_i(\tau(s)) \end{aligned} \tag{6.1.8}$$

is an $\{\mathcal{F}_{\tau(t)}\}$- local martingale for each $f \in D$, and hence, Y is a solution of the martingale problem for $A_2 = \{(f,g) \in A : f \in D,\ B_i f = \beta_i A f \text{ on } E_i,\ i = 1, \ldots, m\}$.

If λ_0 is strictly increasing (cf. Lemma 6.1.6) and $\lim_{u \to \infty} \lambda_0(u) = \infty$ (cf. Lemma 6.1.9), then we can take $\beta_i \equiv 0$, which gives a solution for martingale problems of the form (6.0.2) and leads to the definition of the constrained martingale problem.

6.2 CONSTRAINED MARTINGALE PROBLEMS

As noted above, Example 6.1.3 satisfies the condition of Lemma 6.1.4, Lemma 6.1.6, and Lemma 6.1.9, so λ_0 is strictly increasing and goes to infinity as $t \to \infty$. Define $\tau(t) = \inf\{s : \lambda_0(s)\, t\}$ and $Y(t) = X(\tau(t))$. Then

$$\begin{aligned} f(X(\tau(t))) &- \int_0^{\tau(t)} \frac{1}{2} f''(X(s))d\lambda_0(s) - \int_0^{\tau(t)} f'(X(s))d\lambda_1(s) \\ &= f(Y(t)) - \int_0^{t} \frac{1}{2} f''(Y(s))ds - f'(0) \cdot \lambda_1(\tau(t)) \end{aligned}$$

is a local martingale (actually a martingale) and hence Y is a solution of the martingale problem for $\{(f, \frac{1}{2}f'') : f, f'' \in \hat{C}([0,\infty)),\ f'(0) = 0\}$.

To motivate looking at a more general form for the martingale problem than that given in (6.0.2), modify Example 6.1.3 as follows:

6.2.1 Example Let $E = (-\infty, \infty)$, $E_0 = (0, \infty)$, and $E_1 = (-\infty, 0]$. Let A be the generator for a process with stationary independent increments, for example,

$$Af(x) = \int (f(x+y) - f(x) - yf'(x))\nu(dy) + af'(x) \tag{6.2.1}$$

for $f \in D = C_c^2$, where $\int y \wedge y^2 \nu(dy) < \infty$, and let $Bf = f'$. Let $(X, \lambda_0, \lambda_1)$ be a solution of the patchwork martingale problem for (A, E_0, B, E_1). Then by Remark 1.11,

$\lim_{t\to\infty} \lambda_0(t) = \infty$ (take $f_n(x) = (x + e^x)/(1 + e^x - n^{-1}x)$), but λ_0 is not strictly increasing. We still can define $\tau(t) = \inf\{u : \lambda_0(u) > t\}$ and $Y(t) = X(\tau(t))$. Then

$$\begin{aligned} f(Y(t)) - \int_0^{\tau(t)} Af(X(s))d\lambda_0(s) - \int_0^{\tau(t)} Bf(X(s))d\lambda_1(s) \\ = f(Y(t)) - \int Af(Y(s))ds - \int_0^{\tau(t)} Bf(X(s))d\lambda_1(s) \end{aligned} \tag{6.2.2}$$

is a local martingale, and hence, Y is a solution of the martingale problem for $A_1 = \{(f,g) \in A : Bf = 0 \text{ on } \bar{E}_1\}$. Note that $Bf = 0$ on $\bar{E}_1$ implies that $f(x) = f(0)$ for $x < 0$. Consequently, Y is also a solution of the martingale problem for C given by

$$Cf(x) = \int (f((x+y) \vee 0) - f(x) - yf'(x))\nu(dy) + af'(x).$$

Finally, note that the last term on the right of (6.2.2) can be written as

$$\int_{[0,t]\times \bar{E}_1} Bf(z)\Lambda(ds \times dz)$$

where $\Lambda([0,t] \times F) = \int_0^{\tau(t)} \chi_F(X(s))d\lambda_1(s)$.

Example 6.2.1 illustrates the formulation of the constrained martingale problem we wish to consider. Specifically, suppose we have a solution of a patchwork martingale problem satisfying $\lambda_0(t) \to \infty$ as $t \to \infty$. Let $\tau(t) = \inf\{s : \lambda_0(s) > t\}$, $\mathcal{G}_t = \mathcal{F}_{\tau(t)}$, and set $Y(t) = X(\tau(t))$ and $\Lambda_i\,(t,\Gamma) = \int_0^{\tau(t)} \chi_\Gamma(X(s))\, d\lambda_i(s)$. Then one would expect

$$f(Y(t)) - \int_0^t Af(Y(s))ds - \sum_{i=1}^m \int_{[0,t]\times \bar{E}_i} B_i f(z)\Lambda_i(ds \times dz) \tag{6.2.3}$$

to be at least a $\{\mathcal{G}_t\}$- local martingale. We will say that a process Y is a **solution of the constrained martingale problem for** $(A, E_0, B_1, E_1, \ldots, B_m, E_m)$ if there exists a filtration $\{\mathcal{G}_t\}$ and, for $i = 1, \ldots, m$, a random measure Λ_i on $[0,\infty) \times \bar{\lambda}_i$ such that (6.2.3) is a $\{\mathcal{G}_t\}$-(local) martingale for each $f \in D$.

Unfortunately, Y constructed above need not be a solution of the constrained martingale problem, since $\tau(t)$ is not necessarily continuous and it may not be possible to "stop" the process in such a way that the last term in (6.2.3) has finite expectation (i.e., (6.2.3) may not be a local martingale). Typically, however, this will not be a problem.

Note that the martingale (6.1.4) is bounded by $C_1 + C_2 t$ for some positive constants C_1, C_2. It follows by the optimal sampling theorem (see, for example, Ethier and Kurtz (1986), Theorem 2.2.13) that (6.2.3) is a martingale if $E[\tau(t)]$ is finite. Since $\tau(t) = \sum_{i=1}^{m} \lambda_i(\tau(t)) + t$ it is sufficient to show that $E[\lambda_i(\tau(t))]$ is finite for each i. Noting that η_K as defined in the proof of Lemma 6.1.9 is just $\tau(K)$, (6.1.7) gives the following.

6.2.2 Proposition Under the conditions of Lemma 6.1.9 (or Remark 1.11), $E[\tau(t)] < \infty$ for all $t \geq 0$, and (6.2.3) is a $\{\mathcal{G}_t\}$-martingale for each $f \in D$.

Proof: As in (6.1.7) we have

$$\begin{aligned} E[f(X(n \wedge \tau(t)))] - E[f(X(0))] - E\left[\int_0^{n\wedge\tau(t)} Af(X(s))d\lambda_0(s)\right] \\ \geq E\left[\sum_{i=1}^{m} \int_0^{n\wedge\tau(t)} B_i f(X(s))d\lambda_i(s)\right] \\ \geq \varepsilon E\left[\sum_{i=1}^{m} \lambda_i(n \wedge \tau(t))\right] \end{aligned} \tag{6.2.4}$$

and the desired estimate follows by letting $n \to \infty$.

If τ is continuous, that is, if λ_0 is strictly increasing, then (6.2.3) will be a local martingale without further assumptions.

6.2.3 Proposition Suppose λ_0 is strictly increasing a.s. and $\lim_{t\to\infty} \lambda_0(t) = \infty$. Then $\tau(t)$ is continuous a.s., and for each $f \in D$, (6.2.3) is a $\{\mathcal{G}_t\}$-local martingale and can be written

$$f(Y(t)) - \int_0^t Af(Y(s))ds - \sum_{i=1}^{m} \int_0^t B_i f(Y(s))d\gamma_i(s) \tag{6.2.5}$$

where $\gamma_i(t) = \lambda_i(\tau(t))$. (Note that γ_i is continuous.)

6.2.4 Remark Solutions of the form (6.2.5) will be called **local time solutions** of the constrained martingale problem.

Proof: The continuity of $\tau(t)$ follows immediately from the assumption that λ_0 is strictly increasing. Let

$$\eta_n = \inf\left\{t > 0 : \sum_{i=1}^{m} \lambda_i(\tau(t)) > n\right\}.$$

Then by the continuity of $\tau(t), \sum_{i=1}^{m} \lambda_i(\eta_n) = n$ and $\lim_{n\to\infty} \eta_n = \infty$. In this case, the last term in the stopped process

$$f(Y(t \wedge \eta_n)) - \int_0^{t\ \eta_n} Af(Y(s))ds - \sum_{i=1}^{m} \int_{[0,t\wedge\eta]\times\bar{E}_i} B_i f(z)\Lambda_i(ds \times dz) \qquad (6.2.6)$$

is bounded, so (6.2.6) is a martingale and hence (6.2.5) is a local martingale.

The above proposition shows that a local time solution of a constrained martingale problem can be constructed from a solution of the corresponding patchwork martingale problem provided λ_0 is strictly increasing and tends to infinity. The converse also holds. We say that uniqueness holds for a martingale problem if any two solutions with the same initial distributions have the same finite dimensional distributions.

6.2.5 Corollary There exists a local time solution of the constrained martingale problem for $(A, E_0, B_1, E_1, \ldots, B_m, E_m)$ if and only if there exists a solution of the patchwork martingale problem for $(A, E_0, B_1, E_1, \ldots, B_m, E_m)$ for which λ_0 is strictly increasing and $\lim_{t\to\infty} \lambda_0(t) = \infty$. Uniqueness holds for local time solutions of the constrained martingale problem if and only if uniqueness holds for solutions of the patchwork martingale problem with λ_0 strictly increasing and tending to infinity.

Proof: Suppose $(Y, \gamma_1, \ldots, \gamma_m)$ is a local time solution of the constrained martingale problem. Let $\eta(t) = \inf\{s : s + \gamma_1(s) + \cdots + \gamma_m(s) > t\}$. Define $X = Y \circ \eta$, $\lambda_0 = \eta$, and for $i = 1, \ldots, m$, $\lambda_i = \gamma_i \circ \eta$. Then $(X, \lambda_o, \ldots, \lambda_m)$ is a solution of the patchwork martingale problem by the optional sampling theorem.

We now turn to the question of uniqueness. We consider only the uniqueness of X, since frequently the distribution of X will be uniquely determined but not the distributions of the Λ_i.

6.2.6 Proposition Suppose that for each $\mu \in \wp(\bar{E}_o)$ any two solutions of the constrained martingale problem for $(A, E_0, B_1, E_1, \ldots, B_m, E_m)$ with initial distribution μ have the same one-dimensional distributions. Then each solution of the martingale problem is a Markov process and any two solutions with the same initial distribution have the same finite-dimensional distributions.

Proof: The proof is exactly the same as that of Theorem 4.4.2 of Ethier and Kurtz (1986).

If $\{\lambda f - Af : f \in D,\ B_i f = 0 \text{ on } \bar{E}_i,\ i = 1, \ldots, m\}$ is separating, then conditions for uniqueness are given by Theorem 4.4.1 and Corollary 4.4.4 of Ethier and Kurtz. These results suggest another which more fully exploits the additional structure of the present problem. We say that a collection $\{(g_0, \ldots g_m) : g_i \in B(\bar{E}_i)\}$ is separating if $\mu_i,\ \nu_i \in \mathcal{M}(\bar{E}_i)$, $i = 0, \ldots,\ m$, satisfying

$$\sum_{i=0}^{m} \int_{\bar{E}_i} g_i \, d\mu_i = \sum_{i=0}^{m} \int_{\bar{E}_i} g_i \, d\nu_i$$

implies $\mu_i = \nu_i$, $i = 0, \ldots, m$.

6.2.7 Proposition Suppose that for each $\alpha > 0$ the collection $G_\alpha = \{(g_0, \ldots, g_m) : g_0 = \alpha f - Af \text{ on } \bar{E}_o,\ g_i = B_i f \text{ on } \bar{E}_i,\ i = 1, \ldots, m;\ f \in D\}$ is separating and that there exist $\phi \in D$ and $\varepsilon > 0$ such that $B_i \phi \geq \varepsilon$, $i = 1, \ldots, m$. Then any two solutions of the martingale problem for $(A, E_0, B_1, E_1, \ldots,\ B_m E_m)$ with sample paths in $D_{\bar{E}_0}[0, \infty)$ and the same initial distribution have the same finite dimensional distributions.

Proof: First note that if X is a solution, then for $f \in D$ and $\alpha > 0$

$$f(X(t))e^{-\alpha t} - \int_0^t e^{-\alpha s}(\alpha f(X(s)) - Af(x(s)))ds$$

$$-\sum_{i=1}^{m} \int_{[0,t] \times \bar{E}_i} e^{-\alpha s} B_i f(z) \Lambda_i(ds \times dz)$$

is an $\{\mathcal{F}_t\}$- martingale (the existence of ϕ with the assumed properties assures the needed integrability of the Λ_i). Consequently, if $\tilde{X}$ is also a solution with the same initial distribution we have

$$
\begin{aligned}
E[\int_0^\infty e^{-\alpha s}(\alpha f\,(X(s)) - Af\,(X(s)))ds]& \\
&+\sum_{i=1}^m E[\int_{[0,\infty)\times \bar{E}_i} e^{-\alpha s}\; B_i F(z)\Lambda_i(ds\times dz)] \\
&= -E[f(X(0))] \\
&= -E[f(\tilde{X}(0))] \\
&= E[\int_0^\infty e^{-\alpha s}(\alpha f(\tilde{X}(s)) - Af(\tilde{X}(s)))ds] \\
&+\sum_{i=1}^m E[\int_{[0,\infty)\times \bar{E}_i} e^{-\alpha s} B_i f(z)\tilde{\Lambda}_i(ds\times dz)],
\end{aligned}
$$

and the assumption that G_α is separating implies that

$$
E\left[\int_0^\infty e^{-\alpha s}h(X(s))ds\right] = E\left[\int_0^\infty e^{-\alpha s}h(\tilde{X}(s))ds\right],
$$

for all $h \in B(\bar{E}_0)$. Since $\alpha > 0$ is arbitrary, the uniqueness of the Laplace transform implies that the distribution of $X(t)$ is the same as that of $\tilde{X}(t)$ for almost every $t \geq 0$, and the assumed right continuity implies that equality holds for every t. Uniqueness then follows by Proposition 6.2.6.

6.3 SUBMARTINGALE PROBLEMS

In treating reflecting diffusion processes, Stroock and Varadhan (1971) introduced the notion of a submartingale problem. In our context, we will say that X is a solution of the submartingale problem for $(A, E_0, B_1, E_1, \ldots, B_m, E_m)$ if $X(t) \in \bar{E}_0$ for all $t \geq 0$ a.s. and there exists a filtration $\{\mathcal{F}_t\}$ such that for each $f \in D$ satisfying $B_i f(x) \geq 0$, $\in \bar{E}_i$, $i = 1, \ldots, m$,

$$
f(X(t)) - \int_0^t Af(X(s))ds
$$

is an $\{\mathcal{F}_t\}$-submartingale. Clearly, any solution of the constrained martingale problem for $(A, E_0, B_1, E_1, \ldots, B_m, E_m)$ is a solution of the submartingale problem. As was shown by Stroock and Varadhan for reflecting diffusions and by Anderson (1976) for diffusions with more general boundary conditions, the converse also holds quite generally.

6.3.1 Theorem Suppose there exists $\phi \in D$ and $\epsilon > 0$ such that for $i = 1, \ldots, m$, $B_i\phi \leq \varepsilon$ on $\bar{E}_i$. If X is a solution of the submartingale problem for $(A, E_0, B_1, E_1, \ldots, B_m, E_m)$, then there exist measure valued processes $\{\Lambda_i\}$ such that $(X, \Lambda_1, \ldots, \Lambda_m)$ is a solution of the constrained martingale problem for $(A, E_0, B_1, E_1, \ldots, B_m, E_m)$.

Proof: For each $f \in D$ satisfying $B_i f \geq 0$ on $\bar{E}_i$, the Doob-Meyer decomposition ensures the existence of a nondecreasing predictable process ξ_f

$$f(X(t)) - \int_0^t Af(X(s))ds - \xi_f(t)$$

is an $\{\mathcal{F}_t\}$-martingale. For each $f \in D$, there exists a $c > 0$ such that $B_i(f + c\phi) \geq 0$ on $\bar{E}_i$, for $i = 1, \ldots, m$. We can then define $\xi_f = \xi_{f+c\phi} - c\xi_f$. The uniqueness of the Doob-Meyer decomposition ensures that ξ_f is uniquely determined (up to modification on an event of probability zero), and ξ_f is linear in the sense that

$$\xi_{af_1+bf_2} = a\xi_{f_1} + b\xi_{f_2} \text{ a.s.} \tag{6.3.1}$$

For $f \in D$, define $c_f = \max_i \sup_{x \in \bar{E}_i} |B_i f(x)|/\epsilon$. Then $B_i(c_f\phi + f) \geq 0$ and $B_i(c_f\phi - f) \geq 0$, and it follows that

$$|\xi_f| \leq c_f \xi_\phi \text{ a.s.} \tag{6.3.2}$$

Let L be the space of bounded functions which are continuous on each E_i and whose restrictions to E_i extend to a continuous function on $\bar{E}_i$. Give L the sup norm. Let L_0 be the subspace of L_0 be the subspace of L of functions of the form

$$\sum_{i=1}^{m} \chi_{E_i} B_i f.$$

Fix $t > 0$. Except for the fact that (6.3.1) and (6.3.2) only hold a.s., it follows that the mapping

$$h : \sum_{i=1}^{m} \chi_{E_i} B_i f \to \xi_f(t)$$

defines a bounded linear functional on L_0. Using the separability of L, there exists a countable dense subsapce L_1 of L_0, and using the countability ξ_f can be modified so that (6.3.1) and (6.3.2) hold for all ω on L_1, and by the continuity implied by (6.3.2), the modified ξ_f can be extended to all of L_0. The Hahn-Banach theorem ensures that h can be extended to all of L, and a variant of the Riesz representation theorem implies that for $i = 1, \ldots, m$, there exists a measure Λ_i^t on $\bar{E}_i$ such that

$$\xi_f(t) = \sum_{i=1}^{m} \int_{\bar{E}_i} B_i f(z) \Lambda_i^t(z).$$

The Λ_i^t can be taken to be nondecreasing in t, and defining $\Lambda_i([0,t] \times \Gamma) = \Lambda_i^t(\Gamma)$ determines a measure on $[0, \infty) \times \bar{E}_i$ such that

$$\xi_f(t) = \sum_{i=1}^{m} \int_{[0,t] \times \bar{E}_i} B_i f(z) \Lambda_i^t(ds \times dz).$$

This representation of ξ_f gives the desired result.

6.4 THE PENALTY METHOD

A number of authors (see, for example, Menaldi (1982) and Sznitman and Lions (1984)) have used a "penalty method" to construct reflecting diffusion processes. To see how this approach fits into our context, we will consider the use of the penalty method to obtain a solution of the constrained martingale problem for (A, E_0, B, E_1). Let $\rho \in C(E)$ satisfy $\rho = 0$ on E_0, and $\rho > 0$ on E_0^c. Define $C_n f = Af + n\rho Bf$, $n = 1, 2, \ldots$ Then, assuming $A, B \subset C(E) \times C(E)$, $C_n \subset C(E) \times C(E)$, and for $\nu \in \wp(E_0)$ and each n, there exists a solution Y_n for the martingale problem for C_n such that $Y_n(0)$ has distribution ν (in the reflecting diffusion case, Y_n can be obtained as a solution of an Ito equation). Let

$$\gamma_n(t) = \int_0^t n\rho(Y_n(s))\, ds. \tag{6.4.1}$$

Define

$$\lambda_0^n(t) = \inf\{s : s + \gamma_n(s) > t\}, \qquad \lambda_1^n(t) = t - \lambda_0^n(t), \tag{6.4.2}$$

and set $X_n(t) = Y_n(\lambda_0^n(t))$. Then, setting $\mathcal{F}_t^n = \sigma\{X_n(s) : s \le t\}$,

$$f(X_n(t)) - \int_0^t Af(X_n(s)) d\lambda_0^n(s) - \int_0^t Bf(X_n(s)) d\lambda_1^n(s) \tag{6.4.3}$$

is an $\{\mathcal{F}_t^n\}$-martingale. Note that X_n is not a solution of the patchwork martingale problem for (A, E_0, B, E_1), since λ_0^n does not satisfy (6.1.1) (although λ_1^n does). But,

$$\lambda_0^n(t) = \int_0^t \frac{1}{1 + n\rho(X_n(s))} ds, \tag{6.4.4}$$

and it follows easily that any weak limit point of the sequence $\{\lambda_0^n\}$ will satisfy (6.1.1). Consequently, as in the proof of Lemma 6.1.1, if D is dense in $C(E)$, $\{(X_n, \lambda_0^n, \lambda_1^n)\}$ is relatively compact (in the Skorohod topology) and any weak limit point is a solution of the patchwork martingale problem. For simplicity, assume that the sequence converges in distribution to $(X, \lambda_0, \lambda_1)$.

Let τ_n denote the inverse of λ_0^n (which is strictly increasing), and define τ as in Section 6.2. Then $Y_n = X_n \diamond \tau_n$, and if $\lambda_0(t) \to \infty$, $Y \equiv X \diamond \tau$ is a solution of the constrained martingale problem. If, in addition, λ_0 is strictly increasing, then τ is continuous and $\tau_n \Rightarrow \tau$ in the uniform topology. It follows that $\{Y_n\}$ converges in distribution to Y in the Skorohod topology.

If λ_0 is not strictly increasing, then τ is not continuous and $\{Y_n\}$ may not converge in distribution in the Skorohod topology; however, the set

$$\Gamma = \{t : P\{\tau(t) = \tau(t-),\ X(\tau(t)) = X(\tau(t)-)\} = 1\}$$

contains all but countably many points in $[0, \infty)$, and for $t_1, \ldots, t_k \in \Gamma$, $\{(Y_n(t_1), \ldots, Y_n(t_k))\}$ converges in distribution to $(Y(t_1), \ldots, Y(t_k))$. The nature of this convergence will be explored more fully in Kurtz (1988).

REFERENCES

[1] ANDERSON, R. F. (1976). *Diffusions with second order boundary conditions*, I, II. Indiana Math. J. 25, 367-395, 403-441.

[2] BONY, J.-M., COURRÉGE, P., PRIOURET, P. (1968). *Semi-groupes de feller sur une variété à bord compacte et problèmes aux limites intégro-différentiels du second ordre donnant lieu au principe du maximum.* Ann. Inst. Fourier, Grenoble 18/2, 369-521.

[3] ETHIER, S. N. and KURTZ, T. G. (1986). *Markov Processes: Characterization and Convergence.* Wiley, New York.

[4] IKEDA, N. (1961). *On the construction of two-dimensional diffusion processes satisfying Wentzell's boundary conditions and its application to boundary value problems.* Mem. Coll. Sci. Univ. Kyoto 33, 367-427.

[5] IKEDA, N. and WATANABE, S. (1981). *Stochastic Differential Equations and Diffusion Processes.* North Holland, Amsterdam.

[6] KURTZ, T. G. (1988). *Alternative topologies for convergence in distribution of stochastic processes.* (In preparation.)

[7] LIONS, P. L. and SZNITMAN, A. S. (1984). *Stochastic differential equations with reflecting boundary conditions.* Comm. Pure Appl. Math. 37, 511-537.

[8] MENALDI, J. L. (1982). *Stochastic cotrol problem for reflected diffusions in a convex bounded domain.* Advances in Filtering and Optimal Stochastic Control. Lect. Notes Cont. Inf. Sci. 42, 246-255.

[9] MIKULYAVICHYUS, R. A. (1980a). *On the existence and uniqueness of solutions of the martingale problem on a manifold with boundary.* Lithuanian Math. J. 20, 41-49.

[10] MIKULYAVICHYUS, R. A. (1980b). *On the existence of solutions of the martingale problem on a manifold with boundary.* Lithuanian Math. J. 20, 134-143.

[11] MIKULYAVICHYUS, R. A. (1982). *On the martingale problem.* Russian Math. Surveys 37, 137-150.

[12] SATO, K., and UENO, T. (1965). *Multi-dimensional diffusion and the Markov process on the boundary.* J. Math. Kyoto Univ. 4, 529-605.

[13] SKOROHOD, A. V. (1961). *Stochastic equations for diffusion processes in a bounded region.* Theor. Probab. Appl. 6, 264-274.

[14] STROOCK, D. E. and VARADHAN, S. R. S. (1971). *Diffusion processes with boundary conditions.* Comm. Pure Appl. Math. 24, 147-225.

[15] TAIRA, K. (1977). *Équations aux derivées partielles.* C. R. Acad. Sc. Paris 284, 1133-1136.

[16] TANAKA, H. (1979). *Stochastic differential equations with reflecting boundary condition in convex regions.* Hiroshima Math. J. 9, 163-177.

[17] VENTSEL', A. D. (1959). *On boundary conditions for multidimensional diffusion processes.* Theor. Probab. Appl. 4, 164-177.

[18] WATANABE, S. (1971). *On stochastic differential equations for multi-dimensional diffusion processes with boundary conditions I, II.* J. Math. Kyoto Univ. 11, 169-180, 545-551.

Research supported in part by the National Science Foundation under Grant Number DMS-8401360.

7

APPLICATIONS OF STOCHASTIC CALCULUS IN FINANCIAL ECONOMICS*

Ioannis Karatzas†

ABSTRACT

We shall present a unified approach, based on stochastic calculus, to the problems of *option pricing* and *consumption/investment*, under very general assumptions on the market model.

For the first of these problems, we shall discuss the valuation of both "European"(exercisable only at maturity) and "American"(exercisable any time before or at maturity) contingent claims. Notions and results from the theory of *optimal stopping* will be employed in the treatment of American options.

A general *consumption/investment* problem will also be considered, for an agent whose actions cannot affect the market prices and whose intention is to maximize total expected discounted utility of both consumption and terminal wealth. Under very general conditions on the utility functions of the agent, it will be shown how to approach the above problem by considering separately the two, more elementary ones of maximizing utility from consumption only and of maximizing utility from terminal wealth only, and then appropriately composing them. The optimal consumption and wealth processes will be obtained quite explicitly. In the case of a market model with constant coefficients, the optimal portfolio and consumption rules will be derived very explicitly in feedback form (on the current level of wealth). The Hamilton-Jacobi-Bellman equation of Dynamic Programming associated with this problem, will be reduced to the study of the two *linear* parabolic equations, which will be solved in closed form.

Finally, we shall apply the results on the consumption/investment problem for a single agent, to study the question of *equilibrium* in an economy with several financial agents, whose joint optimal actions determine the price of a traded commodity by "clearing"the markets.

Some familiarity with stochastic calculus, including the fundamental martingale representation and Girsanov theorems, will be assumed. No previous exposure to stochastic control theory will be necessary.

---

*Notes for lectures delivered at the Systems Research Center, University of Maryland, as part of a *Distinguished Lecture Series on Stochastic Calculus.* July 20-31, 1987.

†Supported in part by the National Science Foundation under Grant DMS-84-16736, and by the Shirley Kallek Memorial Fund at Columbia.

7.1 INTRODUCTION AND SUMMARY

Our aim in these notes is to report, to a wider audience than the already well-informed, on certain recent advances in the field of financial economics which have been made possible thanks to the methodologies of stochastic calculus. All the questions treated here are formulated in the context of a financial market which includes a risk-free asset, called the *bond*, and several risky assets called *stocks*; the prices of these latter are driven by an equal number of independent Brownian motions, which model the exogenous forces of uncertainty that influence the market. The interest rate of the bond, the appreciation rates of the stocks as well as their volatilities, constitute the *coefficients* of the market model; we allow them to be arbitrary bounded measurable processes, adapted to the Brownian filtration, but require that a certain nondegeneracy (or "completeness") condition (7.2.3) be satisfied.

The questions that we address include

(i) a general treatment of the *pricing of contingent claims* such as options, both European (exercisable only at maturity) and American (exercisable any time before or at maturity),

(ii) the resolution of *consumption/investment problems* for a "small investor"(i.e., an economic agent whose actions cannot influence the market prices) with quite general utility functions, and

(iii) the associated study of *equilibrium models*; these are formulated in the context of an economy with several small investors and one consumption good, whose price is determined by the joint optimal actions of all these agents in a way that "clears" the markets.

Instrumental in the approach that we adopt are two fundamental results of stochastic calculus: the *Girsanov change of probability measure* and the *representation of Brownian martingales as stochastic integrals*. The former constructs processes that are independent Brownian motions under a new, equivalent probability measure which, roughly speaking, "equates the appreciation rates of all the stocks to the interest rate of the bond." The latter of these results provides the "right portfolios" (investment strategies) for the investors in the abovementioned problems. We assume that the reader is familiar with both these results; they are discussed in several monographs and texts dealing with stochastic calculus, such as Ikeda and

Watanabe (1981) and Karatzas and Shreve (1987).

Here is now an outline of these notes. Sections 7.2 and 7.3 set up the model for the financial market and for the small investor, respectively; this latter has at his disposal the choice of a *portfolio* (investment strategy) and a *consumption strategy*, which determine the evolution of his wealth. The notion of admissible portfolio/consumption strategies, that avoid negative wealth with probability one, is introduced and expounded upon in section 7.5, which can be regarded as the cornerstone of the paper.

Based on the results of section 7.5, we treat the pricing of European contingent claims in section 7.6; we provide the fair price and the subsequent values for such instruments, and derive the celebrated Black and Scholes (1973) formula for European call options as a special case of these results. The analogous problems for American contingent claims are taken up in section 7.12; predictably, their treatment requires notions and results from the theory of optimal stopping.

Sections 7.7-7.11 are concerned with *optimization problems* for a small investor. We introduce the concept of utility function in section 7.7, and treat first a problem in which utility is derived only from consumption (section 7.8); based on the methodology of section 7.5, we provide quite explicit expressions for the optimal consumption and wealth processes, as well as for the associated value $V_1(x)$ of this problem, as a function of the initial wealth $x > 0$. The "dual" situation, with utility derived only from terminal wealth, is discussed in section 7.9; again, explicit expressions are obtained for the abovementioned quantities, including the new value function $V_2(x)$.

We "merge" the two problems in section 7.10, where we take up the more realistic case of utility coming from *both* consumption and terminal wealth; it is shown then that a reasonable "compromise" between the two competing objectives can be achieved in the following fashion. At time $t = 0$, the investor divides his initial capital $x > 0$ into two pieces $x_1 > 0$, $x_2 > 0$, $x_1 + x_2 = x$; for the initial capital x_1 (respectively, x_2) he faces, from then on, a problem in which utility comes only from consumption (respectively, only from terminal wealth). It is shown that this simple procedure, in the form of the superposition of his actions for the two individual problems, yields optimal strategies for the composite problem, provided that x_1, x_2 are chosen so that $V_1'(x_1) = V_2'(x_2)$. Again, explicit expressions are provided for the optimal consumption and wealth processes, and for the resulting value function $V(x) = V_1(x_1) + V_2(x_2)$.

This type of analysis provides no information on the optimal portfolio strategy for the problem of section 7.10, except for guaranteeing its existence. To amend this drawback, we specialize in section 7.11 the problem to the case of *constant coefficients*; we reduce the associated Hamilton-Jacobi-Bellman equation to a system of two *linear* parabolic partial differential equations. With the help of the Feynman-Kac thorem and the Black and Scholes formula, we obtain the solutions of these equations in closed form, and from them the value function $V(x)$ by composition; we also derive very explicit expressions for the optimal portfolio and investment strategies, in feedback form on the current level of wealth.

In section 7.13 we apply the theory of section 7.8 to the study of an *equilibrium model.* We consider an economy with several agents, who can invest in the financial market and receive continuously endowment streams in units of a certain commodity; this latter is traded in the market at a spot price $\psi(\cdot)$. It is shown that the optimal actions of these small investors determine, in principle, the price ψ according to the law of "supply and demand," which mandates that the commodity be consumed entirely as it enters the economy and that the net demand for each financial asset be zero.

7.2 THE MODEL FOR THE FINANCIAL MARKET

We shall deal exclusively in this paper with a financial market in which $d+1$ assets (or "securities") can be traded continuously. One of them is a non-risky asset, called the *bond*, with price-per-share $P_0(t)$ governed by the equation

$$dP_0(t) = P_0(t)r(t)dt \ . \tag{7.2.1}$$

In addition to the bond, there are d risky securities in the market, called *stocks*. The price $P_j(t)$ for one share of the jth stock is modelled by the linear stochastic differential equation

$$dP_j(t) = P_j(t)[b_j(t)dt + \sum_{k=1}^{d} \sigma_{jk}(t)dW_k(t)] \ ; \quad j = 1,2,\ldots,d \ . \tag{7.2.2}$$

In this model we have as many stocks as there are sources of uncertainty; the latter are modelled by the components of the d-dimensional Brownian motion $W(t) =$

$(W_1(t), \ldots, W_d(t))^*$. With this interpretation, $\sigma_{jk}(\cdot)$ in (7.2.2) gives the instantaneous intensity with which the kth source of uncertainty affects the price of the jth stock.

The probabilistic setting will be as follows: the Brownian motion W will be defined on the complete probability space $(\Omega, \mathcal{F}, P)$, and we shall denote by $\{\mathcal{F}_t\}$ the P-augmentation of the natural filtration

$$\mathcal{F}_t^W = \sigma(W(s);\ \ 0 \le s \le t)\ ;\quad 0 \le t < \infty\ .$$

The *interest rate* process $r(t)$, $0 \le t < \infty$ of the bond; the *appreciation rate* vector process $b(t) = (b_1(t), \ldots, b_d(t))^*$, $0 \le t < \infty$ of the stocks; and the *volatility* matrix-valued process $\sigma(t) = \{\sigma_{jk}(t)\}_{1 \le j,k \le d}$, $0 \le t < \infty$; will all be measurable, adapted to $\{\mathcal{F}_t\}$, and bounded uniformly in $(t, \omega) \in [0, \infty) \times \Omega$. We shall also assume that the *covariance matrix* process $a(t) \triangleq \sigma(t)\sigma^*(t)$ is strongly nondegenerate: i.e., there exists a number $\epsilon > 0$ such that

$$\xi^* \sigma(t,\omega)\sigma^*(t,\omega)\xi \ge \epsilon ||\xi||^2;\quad \forall\quad \xi \epsilon \mathcal{R}^d,\ \ (t,\omega) \in [0,\infty) \times \Omega\ . \tag{7.2.3}$$

The assumption (7.2.3) amounts to what is called *completeness of the market model* in the finance literature, and will enable our analysis to go through without serious technical difficulties. In particular, it is not hard to verify that, under (7.2.3), both $\sigma(t, \omega)$ and $\sigma^*(t, \omega)$ are invertible, and that we have

$$||(\sigma(t,\omega))^{-1}\xi|| \le \frac{1}{\sqrt{\epsilon}} ||\xi||\ ;\quad \forall\quad \xi \epsilon \mathcal{R}^d \tag{7.2.4}$$

$$||(\sigma^*(t,\omega))^{-1}\xi|| \le \frac{1}{\sqrt{\epsilon}} ||\xi||\ ;\forall\quad \xi \epsilon \mathcal{R}^d \tag{7.2.5}$$

for every $(t, \omega) \in (0, \infty) \times \Omega$.

7.3 A "SMALL INVESTOR"

Let us consider now an economic agent, who invests in the various securities and whose decisions cannot affect the prices in the market (a "small investor"). we shall denote by $X(t)$ the *wealth* of this agent at time t, by $\pi_j(t)$ the amount that he invests in the jth stock at that time $(1 \le j \le d)$, and by $c(t)$ the rate at which he withdraws money for consumption.

Notice that we allow here any $\pi_j(t),\ 1 \le j \le d$ to become negative, which amounts to selling the jth stock short. Similarly, the quantity

$$X(t) - \sum_{j=1}^{d} \pi_j(t)$$

invested in the bond at any particular time, may also become negative; this is to be interpreted as borrowing at the interest rate $r(t)$.

7.3.1 Definition

A *portfolio process* $\pi(t) = (\pi_1(t), \dots, \pi_d(t))^*\ ;\ 0 \le t < \infty$ is a measurable, $\mathcal{R}^d$-valued process which is adapted to $\{\mathcal{F}_t\}$ and satisfies

$$\int_0^T ||\pi(t)||^2 dt < \infty\ , \qquad \text{a.s.}$$

for every finite number $T > 0$.

7.3.2 Definition

A *consumption rate process* $c(t),\ 0 \le t < \infty$, is measurable, nonnegative, adapted to $\{\mathcal{F}_t\}$, and satisfies

$$\int_0^T c(t)dt < \infty\ , \qquad \text{a.s.} \tag{7.3.1}$$

for every finite number $T > 0$.

The adaptivity condition in Definitions 7.3.1, 7.3.2 means of course that the investor cannot anticipate the future market prices; thus, "insider trading" is excluded!

With the above interpretation and notation, we have the following equation for the wealth $X(t)$ of the agent:

$$\begin{aligned} dX(t) &= \sum_{j=1}^{d} \pi_j(t) \cdot [b_j(t)dt + \sum_{k=1}^{d} \sigma_{jk}(t)dW_k(t)] \\ &+ (X(t) - \sum_{j=1}^{d} \pi_j(t)) \cdot r(t)dt - c(t)dt \\ &= (r(t)X(t) - c(t))dt + \pi^*(t)(b(t) - r(t)\underset{\sim}{1}\, dt + \pi^*(t)\sigma(t)dW(t), \end{aligned} \tag{7.3.2}$$

where $\underset{\sim}{1}$ is the vector in $\mathcal{R}^d$ with all components equal to 1.

Now (7.3.2) is a very simple linear stochastic differential equation for X; it becomes even simpler, though, if $b_j(t) \equiv r(t),\ \forall\ t \in [0,\infty),\ 1 \le j \le d$, i.e., if the interest rate r matches exactly the appreciation rates of all the stocks. In this very special case, the drift $\pi^*(t)(b(t) - r(t)\underset{\sim}{1})$ disappears from the right-hand side of (7.3.2).

More generally, the celebrated Girsanov (1960) theorem will allow us to remove this drift by considering instead of W a new Brownian motion $\tilde{W}$, under a different probability measure $\tilde{P}$ which effectively "equates"all the appreciation rates b_j to the interest rate r of the bond. We do this in the next section.

7.4 AUXILIARY PROBABILITY MEASURE

Let us introduce the process

$$\theta(t) \triangleq (\sigma(t))^{-1}[b(t) - r(t)\underset{\sim}{1}], \tag{7.4.1}$$

and observe from (7.2.4) and our assumptions on b, r that it is bounded, measurable and adapted to $\{\mathcal{F}_t\}$. Then the exponential supermartingale

$$Z_t \triangleq \exp\{-\int_0^t \theta^*(s)dW(s) - \frac{1}{2}\int_0^t ||\theta(s)||^2 ds\},\ \mathcal{F}_t\ ;\quad 0 \le t < \infty \tag{7.4.2}$$

is actually a martingale.

We fix, from now on, a finite time-horizon $[0,T]$ *on which we shall treat all our problems.* We define the probability measure

$$\tilde{P}(A) \triangleq E(Z_T 1_A)\ ;\quad A \epsilon \mathcal{F}_T \tag{7.4.3}$$

on $(\Omega, \mathcal{F}_T)$, $\tilde{P}$ and P are mutally absolutely continuous on $\mathcal{F}_T$, and the process

$$\tilde{W}(t) \triangleq W(t) + \int_0^t \theta(s)ds\ ;\quad 0 \le t \le T \tag{7.4.4}$$

is Brownian motion under $\tilde{P}$ (Girsanov (1960) or Karatzas & Shreve (1987), section 7.3.5).

In terms of this new process $\tilde{W}$, the equations (7.2.2), (7.3.2) are written as

$$dP_j(t) = P_j(t)[r(t)dt + \sum_{k=1}^{d} \sigma_{jk}(t)d\tilde{W}_k(t)] \tag{7.4.5}$$

$$dX(t) = (r(t)X(t) - c(t))dt + \pi^*(t)\sigma(t)d\tilde{W}(t) , \quad (7.4.6)$$

and their solutions are given by

$$P_j(t)e^{-\int_0^t r(u)du} = P_j(0)\exp\{\int_0^t \sigma_j^*(s)d\tilde{W}(s) - \frac{1}{2}\int_0^t ||\sigma_j(s)||^2 ds\} , \quad (7.4.7)$$

$$X(t)e^{-\int_0^t r(u)du} = x - \int_0^t c(s)e^{-\int_0^s r(u)du}ds + \int_0^t e^{-\int_0^s r(u)du}\pi^*(s)\sigma(s)d\tilde{W}(s) , \quad (7.4.8)$$

respectively. Here, $X_0 = x > 0$ is the investor's initial wealth and $\sigma_j(t) \triangleq (\sigma_{j1}(t), \ldots, \sigma_{jd}(t))^*$. In particular, we conclude from (7.4.7) that *the discounted stock price processes* $P_j(t)e^{-\int_0^t r(u)du}$ *are martingales under* $\tilde{P}$; similarly, it follows from (7.4.8) that the process

$$M(t) \triangleq X(t)e^{-\int_0^t r(u)du} + \int_0^t c(s)e^{-\int_0^s r(u)du}ds , \quad (7.4.9)$$

which consists of discounted wealth at t, plus total discounted consumption on $[0,t]$, is a local martingale under $\tilde{P}$.

7.5 ADMISSIBLE STRATEGIES

We shall single out those pairs (π, c) for which the investor avoids going into debt (i.e., negative wealth) with probability one.

7.5.1 Definition

A pair (π, c) of portfolio and consumption rate processes is called *admissible for the initial wealth* $x \geq 0$, if the corresponding wealth process X of (7.4.8) satisfies

$$X(t) \geq 0 ; \quad \forall\ 0 \leq t \leq T$$

almost surely. The class of such pairs is denoted by $\mathcal{A}(x)$.

For every $(\pi, c) \in \mathcal{A}(x)$, the continuous local martingale M of (7.4.9) is nonnegative, and therefore is a supermartingale. Consequently, with $\mathcal{S}_{u,v}$ denoting the class of $\{\mathcal{F}_t\}$-stopping times with values in $[u,v]$, we have by the optional sampling theorem

$$\tilde{E}[X(\tau)e^{-\int_0^\tau r(u)du} + \int_0^\tau c(s)e^{-\int_0^s r(u)du}ds] \leq x \tag{7.5.1}$$

for every $\tau \in \mathcal{S}_{0,T}$.

7.5.2 Definition

For every given number $x \geq 0$, denote by

(i) $\mathcal{C}(x)$ the class of consumption rate processes c which satisfy

$$\tilde{E}\int_0^T c(s)e^{-\int_0^s r(u)du}ds \leq x \ , \tag{7.5.2}$$

and by

(ii) $\mathcal{L}(x)$ the class of nonnegative, $\mathcal{F}_T$-measurable random variables which satisfy

$$\tilde{E}[Be^{-\int_0^T r(u)du}] \leq x \ . \tag{7.5.3}$$

From the inequality (7.5.1) we deduce

$$(\pi, c) \in \mathcal{A}(x) \ \Rightarrow \ c \in \mathcal{C}(x), \ \ X(T) \in \mathcal{L}(x) \ . \tag{7.5.4}$$

The next two results discuss the extent to which the "opposite" implications are true.

7.5.3 Theorem

For every $c \in \mathcal{C}(x)$, there exists a portfolio process π such that $(\pi, c) \in \mathcal{A}(x)$. If, furthermore, c belongs to the class

$$\mathcal{D}(x) \triangleq \{c \in \mathcal{C}(x); \ \ \tilde{E}\int_0^T c(s)e^{-\int_0^s r(u)du}ds = x\} \ , \tag{7.5.5}$$

then the corresponding wealth process X satisfies $X(T) = 0$ a.s., and the process M of (7.4.9) is a $\tilde{P}$-martingale.

Proof: Given $c \in \mathcal{C}(x)$, we introduce the random variable

$$D \triangleq \int_0^T c(s)e^{-\int_0^s r(u)du}ds$$

and the nonnegative process

$$\xi(t) \triangleq \tilde{E}[\int_t^T c(s)e^{-\int_t^s r(u)du}ds|\mathcal{F}_t] + (x - \tilde{E}D)\ e^{\int_0^t r(u)du} \ ; \quad 0 \leq t \leq T \ . \tag{7.5.6}$$

We have $\xi(0) = x$, and with

$$m(t) \triangleq \tilde{E}(D|\mathcal{F}_t) - \tilde{E}D \tag{7.5.7}$$

we can write (7.5.6) as

$$\xi(t)e^{-\int_0^t r(u)du} = x + m(t) - \int_0^t c(s)e^{-\int_0^s r(u)du}ds. \tag{7.5.6$'$}$$

Now using the fundamental martingale representation theorem (e.g. Ikeda & Watanabe (1981), p. 80 or Karatzas & Shreve (1987), Problem 3.4.16), it can be shown that there exists an $\{\mathcal{F}_t\}$-progressively measurable process φ with values in $\mathcal{R}^d$ and $\int_0^T ||\varphi(s)d||^2 ds < \infty$, a.s. for which the martingale of of (7.5.7) takes the form

$$m(t) = \int_0^t \varphi^*(s)d\tilde{W}(s) . \tag{7.5.7$'$}$$

It suffices then to define

$$\pi(t) \triangleq e^{\int_0^t r(u)du}(\sigma^*(t))^{-1}\varphi(t); \tag{7.5.8}$$

thanks to (7.2.5), this is easily seen to be a portfolio process, and from $(7.5.6)'$, $(7.5.7)'$, (7.5.8) and (7.4.8), the process $X \equiv \xi$ is identified with the wealth corresponding to the pair (π, c).

If $c \in \mathcal{D}(x)$, then we have $\xi(T) = 0$ from (7.5.6) and thus $M(T) = \int_0^T c(s)e^{\int_0^s r(u)du}ds$. The $\tilde{P}$-supermartingale M has then constant expectation:

$$\tilde{E}M(T) = \tilde{E}\int_0^T c(s)e^{-\int_0^s r(u)du}ds = x = \tilde{E}M(0),$$

and is, therefore, a martingale.

The following "controllability"result is analogous to Theorem 7.5.3, and characterizes the *levels of terminal wealth attainable by an initial endowment* $X_0 = x$.

7.5.4 Theorem

(i) For every $B \epsilon \mathcal{L}(x)$, there exists a pair $(\pi, c) \in \mathcal{A}(x)$ such that the corresponding wealth process X satisfies $X(T) = B$, a.s.

(ii) If, furthermore, B belongs to the class

$$\mathcal{M}(x) \triangleq \{B \epsilon \mathcal{L}(x);\ \ \tilde{E}[Be^{-\int_0^T r(u)du}] = x\} , \tag{7.5.9}$$

then $c \equiv 0$ in the above pair, and the resulting process $M(t) = X(t)e^{-\int_0^t r(u)du}$ of (5.4.9) is a $\tilde{P}$- martingale.

Proof: Given $B \epsilon \mathcal{L}(x)$, define the nonnegative process η by

$$\eta(t)e^{-\int_0^t r(u)du} \triangleq \tilde{E}(Q|\mathcal{F}_t) + (x - \tilde{E}Q)(1 - \frac{t}{T}) = x + n(t) - \rho t \,, \qquad (7.5.10)$$

where $Q \triangleq Be^{-\int_0^T r(u)du}$, $\rho \triangleq \frac{x - \tilde{E}Q}{T}$, $n(t) \triangleq \tilde{E}(Q|\mathcal{F}_t) - \tilde{E}Q$. The sought consumption rate process is then $c(t) = \rho \, e^{\int_0^t r(u)du}$; the rest of the argument follows the proof of Theorem 7.5.3.

7.5.5 Remark

Theorem 7.5.4 still holds, if T is replaced by an arbitrary stopping time $\tau \epsilon \mathcal{S}_{0,T}$ (recall (7.5.1)). One would have to replace (7.5.10) by

$$\eta(t)e^{-\int_0^t r(u)du} \triangleq \tilde{E}(Q|\mathcal{F}_t) + (x - \tilde{E}Q)(1 - \frac{t \wedge \tau}{\tau}) \qquad (7.5.10)'$$

and take $c(t) \equiv 0$, $\pi(t) \equiv 0$ for $\tau \leq t \leq T$. The rest of the argument goes through without change.

7.6 THE PRICING OF EUROPEAN OPTIONS

Suppose that at time $t = 0$ we sign a contract which gives us the option to buy, at the specified time T (maturity, expiration date), one share of the stock $j = 1$ at a specified price q (the contractual "exercise price"). At maturity, if the price $P_1(T)$ of the share is below the exercise price, the contract is worthless to us; on the other hand, if $P_1(T) > q$, we can exercise our option at $t = T$, buy one share of the stock at the pre-assigned price q, and then sell the share immediately in the market for $P_1(T)$.

Thus, this contract is equivalent to a payment of $(P_1(T) - c)^+$ at maturity; it is called a *European option*, in contradistinction with "American options"which can be exercised at any stopping time in $[0, T]$; cf. section 7.12.

The following definition generalizes the concept of European option.

7.6.1 Definition

A European Contingent Claim (ECC) is a financial instrument consisting of a payment B at maturity; here, B is a nonnegative, $\mathcal{F}_T$-measurable random variable with

$$E(B^\mu) < \infty, \quad \text{for some} \quad \mu > 1 \; . \tag{7.6.1}$$

7.6.2 Remark

Using the boundedness of the process r and θ, as well as the Hölder inequality, it is not hard to see that (7.6.1) implies

$$\tilde{E}[Be^{-\int_0^T r(u)du}] < \infty \; . \tag{7.6.2}$$

7.6.3 Definition

A *hedging strategy* against the ECC of Definition 7.6.1 is a pair $(\pi, c) \in \mathcal{A}(x)$ for some $x > 0$, such that $X(T) = B$ a.s.

We denote by $\mathcal{H}(x)$ the class of hedging strategies with initial wealth $X_0 = x$.

In words, a hedging strategy $(\pi, c) \in \mathcal{H}(x)$ starts out with initial wealth x and "reproduces the payoff from the ECC" at $t = T$.

What is a fair price to pay at $t = 0$ for the ECC? If there exists a hedging strategy for some $x > 0$, then an agent who contemplates buying the ECC at time $t = 0$ can instead invest in the market according to the portfolio π and consume at the rate c, and still achieve the same wealth at $t = T$ as the payment from the ECC. Therefore, the price he should be prepared to pay at $t = 0$ for the ECC cannot possibly be greater than this amount x.

It is natural then to define the fair price as the smallest value of the initial wealth, which allows the construction of a hedging strategy.

7.6.4 Definition

The number

$$v \triangleq \inf\{x > 0; \;\; \exists \; (\pi, c) \in \mathcal{H}(x)\} \tag{7.6.3}$$

is called the *fair price* at $t = 0$ for the ECC of Definition 7.6.1.

7.6.5 Theorem

The fair price v of Definition 7.6.4 is given as

$$v = \tilde{E}[Be^{-\int_0^T r(u)du}] . \qquad (7.6.4)$$

There exists a portfolio process π with $(\pi, 0) \in \mathcal{H}(v)$, and the corresponding wealth process is given by

$$X(t) = \tilde{E}[Be^{-\int_t^T r(u)du}|\mathcal{F}_t] ; \quad 0 \le t \le T . \qquad (7.6.5)$$

Proof: All the claims follow directly from Theorem 7.5.4.

The random variable $X(t)$ of (7.6.5) is called the *value at time* t of the ECC; it can be shown that for any $(\tilde{\pi}, \tilde{c}) \in \mathcal{H}(v)$, the corresponding wealth process $\tilde{X}$ is indistinguishable from X of (7.6.5).

7.6.6 Example

Consider a financial market model with constant interest rate $r(t) \equiv r \ge 0$, and volatility matrix $\sigma(t) \equiv \sigma$, as well as and a contingent claim with $B = \varphi(P(T))$. Here, $\varphi : \mathcal{R}_+^d \to [0, \infty)$ is a continuous function and

$$P(t) = (P_1(t), \ldots, P_d(t))^* \qquad (7.6.6)$$

is the vector of stock price processes which satisfy, in this case, the equations (7.4.5) in the form

$$dP_j(t) = P_j(t)[rdt + \sum_{k=1}^d \sigma_{jk}\, d\tilde{W}_k(t)] ; \quad 1 \le j \le d. \qquad (7.6.7)$$

The solution of these equations is given by (7.4.7), namely

$$P_j(t) = p_j \exp[(r - \frac{1}{2}a_{jj})t + \sum_{k=1}^d \sigma_{jk}\tilde{W}_k(t)] \qquad (7.6.8)$$

with $P_j(0) = p_j$. We introduce now the function

$$h(t, p, y) : [0, \infty) \times \mathcal{R}_+^d \times \mathcal{R}^d \to \mathcal{R}_+^d$$

via

$$h_j(t,p,y) \triangleq p_j \exp[(r - \frac{1}{2}a_{jj})t + y_j] \; ; \quad 1 \le j \le d \, , \tag{7.6.9}$$

and observe that (7.6.8) can be written in the vector form

$$P(t) = h(t, p, \sigma \tilde{W}(t)). \tag{7.6.10}$$

Coming now to the ECC with $B = \varphi(P(T))$, we see from (7.6.5), (7.6.9) that its value is given by

$$\begin{aligned} X(t) &= \tilde{E}[e^{-r(T-t)} \; \varphi(P(T)) | \mathcal{F}_t] \\ &= \tilde{E}[e^{-r(T-t)} \; \varphi(h(T-t, P(t), \sigma(\tilde{W}(T) - \tilde{W}(t)))) | \mathcal{F}_t] \\ &= e^{-r(T-t)} \int_{\mathcal{R}^d} \varphi(h(T-t, P(t), \sigma z)) \; \Gamma_{T-t}(z) \; dz \end{aligned}$$

a.s. $\tilde{P}$, for every $t \in [0,T)$, where

$$\Gamma_t(z) \triangleq (2\pi t)^{-d/2} \exp\{-\frac{||z||^2}{2t}\} \; ; \qquad z \epsilon \mathcal{R}^d \, , \quad t > 0$$

is the fundamental Gaussian kernel. It follows that, with

$$v(t,p) \triangleq \begin{cases} e^{-r(T-t)} \int_{\mathcal{R}^d} \varphi(h(T-t,p,\sigma z) \; \Gamma_{T-t}(z) \; dz & ; 0 \le t < T, \quad p \in \mathcal{R}^d_+ \\ \varphi(p) & ; t = T, \quad p \in \mathcal{R}^d_+, \end{cases} \tag{7.6.11}$$

the value at time t of the ECC is given by

$$X(t) = v(t, P(t)). \tag{7.6.12}$$

In this case it is even possible to "compute"the portfolio $\pi(t)$ that achieves the value process of (7.6.12). Indeed, under appropriate growth conditions on φ, the function $v(t,p)$ of (7.6.11) is the unique solution of the Cauchy problem

$$\frac{\partial v}{\partial t} + \frac{1}{2} \sum_{j=1}^{d} \sum_{k=1}^{d} a_{jk} \; p_j p_k \; \frac{\partial^2 v}{\partial p_j \partial p_k} + \sum_{j=1}^{d} r p_j \; \frac{\partial v}{\partial p_j} - rv = 0 \; ; \text{ on } \; [0,T) \times \mathcal{R}^d_+$$

$$v(T,p) = \varphi(p) \; ; \qquad p \in \mathcal{R}^d_+$$

by the Feynman-Kac theorem. Applying Itô's rule to the process X of (7.6.12) and using the above equation and (7.6.7), one arrives at

$$dX(t) = rX(t)dt + \sum_{j=1}^{d} \sum_{k=1}^{d} \sigma_{jk} P_j(t) \frac{\partial}{\partial p_j} v(t, P(t)) d\tilde{W}_k(t).$$

A comparison with (7.4.6) gives then

$$\pi_j(t) = P_j(t) \cdot \frac{\partial}{\partial p_j} v(t, P(t)) \; ; \quad 0 \le t \le T, \; 1 \le j \le d$$

for the portfolio process of Theorem 7.6.5. In other words, one should hold $N_j(t) = \frac{\partial}{\partial p_j} v(t, P(t))$ shares of the jth stock at time t.

7.6.7 Remark

In the particular case of a European option as in Example 7.6.6, with $d = 1$, $\varphi(p) = (p - q)^+$ and exercise price $q > 0$, the integration in (7.6.11) can be carried out in a somewhat more explicit form. Indeed, with $\Phi(z) \triangleq \frac{1}{\sqrt{2\pi}} \int_{-\infty}^{z} \exp(-\frac{x^2}{2}) dx$ and

$$\nu_{\pm}(t, p; q) \triangleq \frac{1}{\sigma_{11}\sqrt{t}} [\log(\frac{p}{q}) + (r \pm \frac{1}{2}\sigma_{11}^2)t], \qquad \text{we have}$$

$$v(t, p; q) = \begin{cases} p\Phi(\nu_+(T-t, p; q)) - qe^{-r(T-t)}\Phi(\nu_-(T-t, p; q)) & ; 0 \le t < T, 0 < p < \infty \\ (p-q)^+ & ; t = T, 0 < p < \infty. \end{cases} \tag{7.6.13}$$

Together with

$$X(t; q) \triangleq \tilde{E}[e^{-r(T-t)}(P_1(T) - q)^+ | \mathcal{F}_t] = v(t, P_1(t); q) \; ; \qquad 0 \le t \le T \, , \tag{7.6.14}$$

the expression (7.6.13) constitutes the celebrated *Black & Scholes (1973) formula.* Note that in this formula, as well as in (7.6.11), the appreciation rates of the stocks do not appear; this fact makes these formulas particularly useful, in situations where the appreciation rates are unknown or unobservable.

More generally, any convex and piecewise C^2 function $h : [0, \infty) \to [0, \infty)$ with $h(0) = h'(0) = 0$ can be represented as

$$h(p) = \int_0^{\infty} (p - q)^+ \, h''(q) dq \; .$$

For a ECC with $B = h(P_1(T))$, the value at time t is given then by (7.6.5) as

$$X(t) = \tilde{E}[e^{-r(T-t)} h(P_1(T)) | \mathcal{F}_t] = \int_0^{\infty} h''(q) v(t, P_1(t); q) dq \, , \qquad \text{a.s.}$$

for every $0 \le t \le T$, thanks to the Fubini theorem and (7.6.14).

7.6.8 Remark

If the expiration date T is replaced by a stopping time $\tau \epsilon \mathcal{S}_{0,T}$ and the payment B is an $\mathcal{F}_\tau$ -measurable random variable, the theory of this section still goes through with minor changes. A hedging strategy (Definition 7.6.3) now has to satisfy $X(\tau) = B$ a.s., and (7.6.4), (7.6.5) become

$$v^{(\tau)} = \tilde{E}[Be^{-\int_0^\tau r(u)du}] \tag{7.6.4$'$}$$

$$X^{(\tau)}(t) = \tilde{E}[Be^{-\int_t^\tau r(u)du}|\mathcal{F}_t] \; , \tag{7.6.5$'$}$$

respectively (recall Remark 7.5.5).

7.7 UTILITY FUNCTIONS

In order to formulate meaningful optimization problems for the small investor of section 7.3, we need the concept of utility function.

Let $U : (0,\infty) \to \mathcal{R}$ be a strictly increasing, strictly concave and continuously differentiable function, with $U(0) \triangleq \lim_{c \downarrow 0} U(c) \geq -\infty$, $U'(0) \triangleq \lim_{c \downarrow 0} U'(c) \leq \infty$ and $U'(\infty) \triangleq \lim_{c \to \infty} U'(c) = 0$. A function with these properties will be called a *utility function.*

We shall denote by $I : [0, U'(0)] \stackrel{onto}{\longrightarrow} [0,\infty]$ the inverse of the strictly decreasing function $U' : [0,\infty] \stackrel{onto}{\longrightarrow} [0, U'(0)]$, and extend it by continuity on the entire of $[0,\infty]$, i.e., by setting $I(y) = 0$ for $U'(0) \leq y \leq \infty$. The inequality

$$U(I(y)) \geq U(c) + yI(y) - yc; \quad \forall \; c \,\epsilon\, [0,\infty), \; y \,\epsilon\, (0,\infty) \tag{7.7.1}$$

is then easily verified; we shall find it particularly helpful in comparison arguments.

For certain of our results we shall need to impose the condition

$$U \,\epsilon\, C^2 \quad \text{and} \quad U'' \quad \text{is nondecreasing on} \quad (0,\infty) \, . \tag{7.7.2}$$

Under (7.7.2), I is convex on $(0,\infty)$ and of class C^1 on $(0,\infty)\backslash)(U'(0)) = 0$, we obtain the identit

$$(U(I(y)))' = yI'(y); \quad \forall \; y \,\epsilon$$

7.8 MAXIMIZATION OF UTILITY FROM CONSUMPTION

In this section we shall try to address the following question: how should a small investor, endowed with initial wealth $x > 0$, choose at every time his stock portfolio $\pi(t)$ and his consumption rate $c(t)$, in order to obtain a maximum utility from consumption and at the same time remain solvent, over the horizon $[0,T]$?

In order to give a precise meaning to this question, let us introduce a bounded, measurable and adapted *discount rate* process $\{\beta(t), \mathcal{F}_t; 0 \le t \le T\}$, consider a utility function U_1, and try to maximise the *expected discounted utility from consumption*

$$J_1(x;\pi,c) \triangleq E\int_0^T e^{-\int_0^t \beta(s)ds}\, U_1(c(t))dt \tag{7.8.1}$$

over the class $\mathcal{A}_1(x)$ of pairs $(\pi,c) \in \mathcal{A}(x)$ which satisfy

$$E\int_0^T e^{-\int_0^t \beta(s)ds}\, U_1^-(c(t))dt < \infty\,. \tag{7.8.2}$$

Because in this problem utility comes only from consumption, and can only increase if the consumption increases, it is conceivable that one should increase the net effect of consumption up to the admissible limit mandated by (7.5.2), by considering consumption rate processes only from the class $\mathcal{D}(x)$ of (7.5.5). In other words, it is easy to show that the value function

$$V_1(x) \triangleq \sup_{(\pi,c)\,\in\,\mathcal{A}_1(x)} J_1(x;\pi,c) \tag{7.8.3}$$

for this problem is expressible as

$$V_1(x) = \sup_{\substack{(\pi,c)\,\in\,\mathcal{A}_1(x)\\ c\,\in\,\mathcal{D}(x)}} J_1(x;\pi,c)\,. \tag{7.8.4}$$

To recapitulate: we have to maximize the expression $E\int_0^T e^{-\int_0^t \beta(s)ds}\, U_1(c(t))dt$, over consumption rate processes c that satisfy

$$\tilde{E}\int_0^T e^{-\int_0^t r(s)ds}c(t)dt = E\int_0^T Z_t\, e^{-\int_0^t r(s)ds}c(t)dt \le x\,.$$

But this constrained maximization problem is not hard: it amounts to making

$\exp\{-\int_0^t \beta(s)ds\} \cdot U_1'(c(t))$ proportional to $Z_t \exp\{-\int_0^t r(s)ds\}$, i.e.,

$$c(t) = I_1(y \cdot Z_t \, e^{\int_0^t (\beta(s)-r(s))ds}); \quad 0 \le t \le T \tag{7.8.5}$$

in the notation $I_1 = (U_1')^{-1}$ of section 7.7. The proportionality constant $y > 0$ has then to be chosen, so that the consumption rate process c of (7.8.5) belongs to $\mathcal{D}(x)$.

We do this by introducing the process

$$\zeta_t \triangleq Z_t \, e^{\int_0^t (\beta(s)-r(s))ds} \tag{7.8.6}$$

and the nonnegative, decreasing function

$$\begin{aligned} \mathcal{X}_1(y) &\triangleq \tilde{E} \int_0^T e^{-\int_0^t r(s)ds} I_1(y\zeta_t)dt \\ &= E \int_0^T \zeta_t e^{-\int_0^t \beta(s)ds} I_1(y\zeta_t)dt; \quad 0 < y < \infty . \end{aligned} \tag{7.8.7}$$

It is not hard to see that the process of (7.8.6) satisfies the *linear* stochastic equation

$$d\zeta_t = \zeta_t(\beta(t) - r(t))dt - \zeta_t\theta^*(t)dW(t) . \tag{7.8.8}$$

On the other hand, under the assumption

$$\mathcal{X}_1(y) < \infty, \qquad \forall \; 0 < y < \infty \tag{7.8.9}$$

and with

$$\bar{y}_1 \triangleq \sup\{y > 0; \; \mathcal{X}_1 \text{ is } \textit{strictly} \text{ decreasing on } (0, y)\} ,$$

the function $\mathcal{X}_1$ is continuous and strictly decreasing on $(0, \bar{y}_1)$, and satisfies $\mathcal{X}_1(0) \triangleq \lim_{y \downarrow 0} \mathcal{X}_1(y) = \infty$, $\mathcal{X}_1(\bar{y}_1) = 0$.

If we denote by $\mathcal{Y}_1 : [0, \infty] \xrightarrow{onto} [0, \bar{y}_1]$ the (continouous, strictly decreasing) inverse of the function $\mathcal{X}_1$ and fix a number $x = x_1 \in (0, \infty)$, then the process of (7.8.5) will belong to $\mathcal{D}(x_1)$ for the choice $y = \mathcal{Y}_1(x_1)$; consequently,

$$c_1(t) \triangleq I_1(\mathcal{Y}_1(x_1)\zeta_t) \; ; \quad 0 \le t \le T \tag{7.8.10}$$

will be our candidate optimal consumption rate process. From Theorem 7.5.3, there exists a portfolio process π_1 such that $(\pi_1, c_1) \in \mathcal{A}(x_1)$ and the wealth process X_1

corresponding to this pair is given by

$$X_1(t)e^{-\int_0^t r(s)ds} = \tilde{E}[\int_t^T c_1(s)e^{-\int_0^s r(u)du}\, ds|\mathcal{F}_t]$$

$$= x_1 - \int_0^t c_1(s)e^{-\int_0^s r(u)du}\, ds + \int_0^t e^{-\int_0^s r(u)du}\pi_1^*(s)\sigma(s)d\tilde{W}(s)\,. \tag{7.8.11}$$

This is a nonnegative process which vanishes at $t = T$; if $U_1'(0) = \infty$, then X_1 is positive on $[0, T)$.

We can now use the fundamental inequality (7.7.1) to show that the process c_1 of (7.8.10) satisfies the finiteness condition (7.8.2), and that

$$E\int_0^T e^{-\int_0^t \beta(s)ds}U_1(c(t))dt \le E\int_0^T e^{-\int_0^t \beta(s)ds}U_1(c_1(t))dt \tag{7.8.12}$$

holds for every other process $c \in \mathcal{C}(x_1)$ satisfying (7.8.2). The following result is then established.

7.8.1 Theorem

For a fixed $x_1 > 0$ and with c_1 given by (7.8.10), the above pair (π_1, c_1) belongs to $\mathcal{A}_1(x_1)$ and is optimal for the problem (7.8.1)-(7.8.3):

$$V_1(x_1) = E\int_0^T e^{-\int_0^t \beta(s)ds}\, U_1(c_1(t))dt\,. \tag{7.8.13}$$

Can we characterize the value function of (7.8.3) more precisely? In order to do that, we consider the expected discounted utility

$$G_1(y) \triangleq E\int_0^T e^{-\int_0^t \beta(s)ds}\, U_1(I_1(y\zeta_t))dt; \quad 0 < y < \infty \tag{7.8.14}$$

associated with a consumption process of the form (7.8.5), and assume that

$$E\int_0^T e^{-\int_0^t \beta(s)ds}|U_1(I_1(y\zeta_t))|dt < \infty; \quad \forall\ 0 < y < \infty\,. \tag{7.8.15}$$

Then G_1 is a continuous, decreasing function; moreover, formal differentiation with respect to y in (7.8.7), (7.8.14) under the integral sign, gives

$$\mathcal{X}_1'(y) = E\int_0^T e^{-\int_0^t \beta(s)ds}\zeta_t^2 I_1'(y\zeta_t)dt$$

and

$$G_1'(y) = E\int_0^T e^{-\int_0^t \beta(s)ds}\zeta_t \cdot U_1'(I_1(y\zeta_t))I_1'(y\zeta_t)dt = y\mathcal{X}_1'(y) \ .$$

These formalities can be made rigorous under the condition (7.7.2), and its consequence (7.7.3). One arrives then at the following characterization.

7.8.2 Proposition

Suppose that the utility function U_1 satisfies the conditions (7.7.2), (7.8.9) and (7.8.15). Then the function G_1 of (7.8.14) is decreasing, and differentiable, and satisfies $G_1'(y) = y\mathcal{X}_1'(y); \ \forall \ 0 < y < \infty.$

The value function V_1 of (7.8.3) is given then as

$$V_1 = G_1 \circ \mathcal{Y}_1 \ , \tag{7.8.16}$$

and we have

$$V_1' = \mathcal{Y}_1 \ ; \tag{7.8.17}$$

in particular, V_1 is strictly increasing and strictly concave.

In the important special case $U_1(c) = \log c$, we obtain

$$\mathcal{X}_1(y) = \frac{\alpha_1}{y} \ , \quad G_1(y) = -\alpha_1 \cdot \log y + \delta_1 \tag{7.8.18}$$

and hence

$$V_1(x) = \alpha_1 \cdot \log(\frac{x}{\alpha_1}) + \delta_1$$

with

$$\alpha_1 \triangleq E\int_0^T e^{-\int_0^t \beta(s)ds}dt \qquad \text{and}$$

$$\delta_1 \triangleq E\int_0^T e^{-\int_0^t \beta(s)ds}\{\int_0^t \theta^*(s)dW(s) - \int_0^t (\beta(s) - r(s) - \frac{1}{2}||\theta(s)||^2)ds\}dt \ . \tag{7.8.19}$$

In particular, (7.8.9) and (7.8.15) are then satisfied trivially.

7.8.3 Remark

If $U_1(0) > -\infty$, we have $\mathcal{A}_1(x) = \mathcal{A}(x)$, and (7.8.15) implies (7.8.9).

7.9 MAXIMIZATION OF UTILITY FROM TERMINAL WEALTH

Let us take up now the complementary problem to that of section 7.8, namely the maximization of the *expected discounted utility from terminal wealth*

$$J_2(x;\pi,c) \triangleq E[U_2(X(T))e^{-\int_0^T \beta(s)ds}] \tag{7.9.1}$$

over the class $\mathcal{A}_2(x)$ of pairs $(\pi,c) \in \mathcal{A}(x)$, for which the corresponding wealth processes X satisfy the condition

$$E[U_2^-(X(T))e^{-\int_0^T \beta(s)ds}] < \infty . \tag{7.9.2}$$

Here, U_2 is a given utility function.

In this setting the agent obviously strives to maximize his terminal wealth, within the constraints imposed by the level of his initial endowment and quantified by (7.5.3). Then according to Theorem 7.5.4, he should not consume at all, and should try to achieve a terminal wealth X_T in the class $\mathcal{M}(x)$ of (7.5.9). More precisely, it is again straightforward to show that the value function

$$V_2(x) \triangleq \sup_{(\pi,x)\,\in\,\mathcal{A}_2(x)} J_2(x;\pi,c) \tag{7.9.3}$$

for this problem is given as

$$V_2 = \sup_{\substack{(\pi,0)\,\in\,\mathcal{A}_2(x)\\ X_T\,\in\,\mathcal{M}(x)}} J_2(x;\pi,c) . \tag{7.9.4}$$

The situation is completely analogous to that of the previous section, so we just outline the results. One introduces the decreasing function

$$\mathcal{X}_2(y) \triangleq \tilde{E}[I_2(y\zeta_T)\, e^{-\int_0^T r(s)ds}] ; \qquad 0 < y < \infty , \tag{7.9.5}$$

assumes that

$$\mathcal{X}_2(y) < \infty ; \qquad \forall \;\; 0 < y < \infty , \tag{7.9.6}$$

and notices that $\mathcal{X}_2$ is continuous and strictly decreasing on $(0,\bar{y}_2)$ and that $\mathcal{X}_2(0) \triangleq \lim_{y\downarrow 0} \mathcal{X}_2(y) = \infty$, $\mathcal{X}_2(\bar{y}_2) = 0$ hold, with

$$\bar{y}_2 \triangleq \sup\{y > 0 \; ; \; \mathcal{X}_2 \text{ is } \textit{strictly} \text{ decreasing on } (0,y)\}.$$

Let $\mathcal{Y}_2 : [0,\infty] \xrightarrow{onto} [0, \bar{y}_2]$ denote the inverse of $\mathcal{X}_2$, and fix a number $x = x_2 \in (0,\infty)$; then the random variable

$$X_2(T) \triangleq I_2(\mathcal{Y}_2(x_2)\zeta_T) \tag{7.9.7}$$

belongs to $\mathcal{M}(x_2)$, and is our candidate for the optimal level of terminal wealth. From Theorem 7.5.4, there exists a portfolio process π_2 such that $(\pi_2, 0) \in \mathcal{A}(x_2)$, and the wealth process X_2 corresponding to this pair is given by

$$\begin{aligned} X_2(t)e^{-\int_0^t r(s)ds} &= \tilde{E}[X_2(T)e^{-\int_0^T r(s)ds}|\mathcal{F}_t] \\ &= x_2 + \int_0^t e^{-\int_0^s r(u)du}\pi_2{}^*(s)\sigma(s)d\tilde{W}(s) \ . \end{aligned} \tag{7.9.8}$$

7.9.1 Theorem

For a fixed $x_2 > 0$ and with $X_2(T)$ given by (7.9.7), the above pair $(\pi_2, 0)$ belongs to $\mathcal{A}_2(x_2)$ and is optimal for the problem (7.9.1) - (7.9.3):

$$V_2(x_2) = E[U_2(X_2(T))e^{-\int_0^T \beta(s)ds}] \ . \tag{7.9.9}$$

Sketch of Proof. Using the inequality (7.4.1), one shows first that $X_2(T)$ satisfies (7.9.2), and that

$$E[U_2(X(T))e^{-\int_0^T \beta(s)ds}] \le E[U_2(X_2(T))e^{-\int_0^T \beta(s)ds}] \tag{7.9.10}$$

holds for every random variable $X(T) \in \mathcal{L}(x_2)$ satisfying (7.9.2).

We also have the following characterization of the value function.

7.9.2 Proposition

Under the conditions (7.7.2), (7.9.6) and

$$E[|U_2(I_2(y\zeta_T))|e^{-\int_0^T \beta(s)ds}] < \infty \ ; \qquad \forall \ 0 < y < \infty \tag{7.9.11}$$

on the utility function U_2, the function

$$G_2(y) \triangleq E[U_2(I_2(y\zeta_T))e^{-\int_0^T \beta(s)ds}] \ ; \quad 0 < y < \infty \tag{7.9.12}$$

is decreasing and differentiable, and satisfies $G_2'(y) = y\mathcal{X}_2'(y); \ \forall \ 0 < y < \infty$.

The value function V_2 of (7.9.3) is then given as the composition

$$V_2 = G_2 \circ \mathcal{Y}_2 \ , \tag{7.9.13}$$

and we have

$$V_2' = \mathcal{Y}_2 \ ; \tag{7.9.14}$$

in particular, V_2 is strictly increasing and strictly concave.

In the special case $U_2(c) = \log c$ we deduce easily

$$\begin{aligned} \mathcal{X}_2(y) &= \frac{\alpha_2}{y}\ , \quad G_2(y) = -\alpha_2 \cdot \log y + \delta_1 \qquad \text{and} \\ V_2(x) &= \alpha_2 \cdot \log(\frac{x}{\alpha_2}) + \delta_2\ , \end{aligned} \tag{7.9.15}$$

where

$$\begin{aligned} \alpha_2 &\triangleq Ee^{-\int_0^T \beta(s)ds} \\ \delta_2 &\triangleq E[e^{-\int_0^T \beta(s)ds}\{\int_0^T \theta^*(s)dW(s) - \int_0^T (\beta(s) - r(s) - \frac{1}{2}||\theta(s)||^2)ds\}]\ . \end{aligned} \tag{7.9.16}$$

7.10 MAXIMIZATION OF UTILITY FROM BOTH CONSUMPTION AND TERMINAL WEALTH

Let us consider now an investor who derives utility both from "living well"(i.e., from consumption) and from "becoming rich"(i.e., from terminal wealth). His *expected total discounted utility* is then

$$\begin{aligned} J(x;\pi,c) &\triangleq J_1(x;\pi,c) + J_2(x;\pi,c) \\ &= E\int_0^T e^{-\int_0^t \beta(s)ds} U_1(c(t))dt + E[U_2(X(T))e^{-\int_0^T \beta(s)ds}], \end{aligned} \tag{7.10.1}$$

and he tries to maximize $J(x;\pi,c)$ over $\mathcal{A}_{1,2}(x) \triangleq \mathcal{A}_1(x) \cap \mathcal{A}_2(x)$:

$$V(x) \triangleq \sup_{(\pi,x)\,\epsilon\,\mathcal{A}_{1,2}(x)} J(x;\pi,c)\ . \tag{7.10.2}$$

In contrast to the problems of sections 7.8 and 7.9, this one calls for balancing *competing objectives.* We shall show that the right compromise can be drawn in a very simple manner: at time $t = 0$, the investor just divides his endowment x into two nonnegative parts x_1 and x_2, with $x_1 + x_2 = x$. For x_1, he solves the problem of section 7.8 (with utility U_1 from consumption); for x_2, he solves the problem of section 7.9 (with utility U_2 from terminal wealth). The superposition of his actions for these two problems will lead to the optimal policy for the problem of (7.10.2), provided x_1 and x_2 are chosen judiciously. We shall show how this can be done (cf. (7.10.10) below).

For concreteness, we assume throughout this section that the value functions U_1, U_2 satisfy (7.7.2), as well as the requirements (7.8.9), (7.8.15) and (7.9.6), (7.9.11).

Let us start with an arbitrary pair $(\pi, x) \in \mathcal{A}_{1,2}(x)$ and define

$$x_1 \triangleq \tilde{E} \int_0^T c(t) e^{-\int_0^t r(s)ds} dt, \quad x_2 \triangleq x - x_1 . \tag{7.10.3}$$

Denoting by X the wealth process corresponding to this pair, we conclude from (7.5.1), (7.10.3) that

$$c \in \mathcal{D}(x_1), \qquad X(T) \in \mathcal{L}(x_2). \tag{7.10.4}$$

Theorem 7.8.1 gives us a pair $(\pi_1, c_1) \in \mathcal{A}_1(x_1)$ which is optimal for $V_1(x_1)$, with corresponding wealth process X_1 satisfying $X_1(T) = 0$, a.s. On the other hand, Theorem 7.9.1 provides a pair $(\pi_2, 0) \in \mathcal{A}_2(x_2)$ which is optimal for $V_2(x_2)$, with corresponding wealth process X_2. If we define now

$$\tilde{\pi} \triangleq \pi_1 + \pi_2, \qquad \tilde{c} \triangleq c_1 \qquad \text{and} \qquad \tilde{X} \triangleq X_1 + X_2 \tag{7.10.5}$$

and add (7.8.11), (7.9.8) memberwise, we obtain

$$\begin{aligned}\tilde{X}(t) e^{-\int_0^t r(s)ds} &= \tilde{E}\left[\int_t^T \tilde{c}(s) e^{-\int_0^s r(u)du} ds + \tilde{X}(T) e^{-\int_0^T r(s)ds} \middle| \mathcal{F}_t\right] \\ &= x - \int_0^t \tilde{c}(s) e^{-\int_0^s r(u)du} ds + \int_0^t e^{-\int_0^s r(u)du} (\tilde{\pi}(s))^* \sigma(s) d\tilde{W}(s) .\end{aligned} \tag{7.10.6}$$

In words, $\tilde{X}$ *is the wealth process corresponding to the pair* $(\tilde{\pi}, \tilde{c}) \in \mathcal{A}_{1,2}(x)$.

But now recall (7.10.4), and add up (7.8.12), (7.9.10) memberwise to obtain

$$J(x; \pi, c) \le V_1(x_1) + V_2(x_2) ,$$

whence

$$V(x) \le V_*(x) \triangleq \max_{\substack{x_1 \ge 0, x_2 \ge 0 \\ x_1 + x_2 = x}} [V_1(x_1) + V_2(x_2)] . \tag{7.10.7}$$

The question now is to find x_1, x_2 for which this maximum is achieved, because then the total expected discounted utility corresponding to the pair $(\tilde{\pi}, \tilde{c})$ of (7.10.5) will be *exactly* equal to $V_*(x)$; this will in turn imply

$$V(x) \equiv V_*(x) \tag{7.10.8}$$

from (7.10.7), and thus the above mentioned pair will be shown to be optimal for the problem of (7.10.2).

But the maximization in (7.10.7) is easy: it amounts to selecting x_1, x_2 so that $V_1'(x_1) = V_2'(x_2)$ or, thanks to (7.8.17), (7.9.14):

$$\mathcal{Y}_1'(x_1) = \mathcal{Y}_2'(x_2) = \lambda \Leftrightarrow x_1 = \mathcal{X}_1(\lambda),\ x_2 = \mathcal{X}_2(\lambda)$$

The constant λ is determined uniquely as follows: we introduce the function

$$\mathcal{X}(y) \triangleq \mathcal{X}_1(y) + \mathcal{X}_2(y) = \tilde{E}[\int_0^T e^{-\int_0^t r(u)du} I_1(y\zeta_t)dt + e^{-\int_0^T r(u)du} I_2(y\zeta_T)] \tag{7.10.9}$$

on $(0, \infty)$, which is continuous and strictly decreasing on $(0, \bar{y})$ with

$$\bar{y} \triangleq \bar{y}_1 \vee \bar{y}_2,\ \lim_{y \downarrow 0} \mathcal{X}(y) = \infty, \quad \mathcal{X}(\bar{y}) = 0$$

. Let $\mathcal{Y}:\ [0, \infty] \xrightarrow{onto} [0, \bar{y}]$ be the inverse of $\mathcal{X}$; then $\lambda = \mathcal{Y}(x)$, and the "optimal partition" of the initial wealth is given by

$$x_1 = \mathcal{X}_1(\mathcal{Y}(x)) , \qquad x_2 = \mathcal{X}_2(\mathcal{Y}(x)) . \tag{7.10.10}$$

If we also introduce the function

$$\begin{aligned} G(y) &\triangleq G_1(y) + G_2(y) \\ &= E[\int_0^T e^{-\int_0^t \beta(s)ds} U_1(I_1(y\zeta_t))dt + e^{-\int_0^T \beta(s)ds} U_2(I_2(y\zeta_T))] , \end{aligned} \tag{7.10.11}$$

which is continuous and decreasing on $(0, \infty)$, it is easy to see from (7.8.16), (7.9.13) that

$$V_*(x) = G(\mathcal{Y}(x)) . \tag{7.10.12}$$

We have established the following result.

7.10.1 Theorem

Under the conditions of this section, the value function V of (7.10.2) is given as

$$V = G \circ \mathcal{Y} . \tag{7.10.13}$$

For a fixed initial endowment $x > 0$, the optimal consumption rate process and the optimal level of terminal wealth are given by

$$\hat{c}(t) = I_1(\mathcal{Y}(x)\zeta_t) \; ; \quad 0 \le t \le T \qquad \text{and} \qquad \hat{X}(T) \triangleq I_2(\mathcal{Y}(x)\zeta_T) \, , \tag{7.10.14}$$

respectively; there exists a portfolio process $\hat{\pi}$ such that $(\hat{\pi}, \hat{c})$ is optimal in $\mathcal{A}_{1,2}(x)$ for (7.10.2), and the corresponding wealth process is

$$\hat{X}(t) = \tilde{E}[\int_t^T e^{-\int_t^s r(u)du} I_1(\mathcal{Y}(x)\zeta_s)ds + e^{-\int_t^T r(u)du} I_2(\mathcal{Y}(x)\zeta_T)|\mathcal{F}_t] \tag{7.10.15}$$

almost surely, for every $0 \le t \le T$.

Notice that the process M of (7.4.9), corresponding to the pair $(\hat{\pi}, \hat{c})$ of Theorem 7.10.1, takes the form

$$\hat{M}(t) = \tilde{E}[\int_0^T e^{-\int_0^s r(u)du} \; I_1(\mathcal{Y}(x)\zeta_s)ds + e^{-\int_0^T r(u)du} \; I_2(\mathcal{Y}(x)\zeta_T)|\mathcal{F}_t] \; ;$$

in particular, it is a $\tilde{P}$-martingale.

7.10.2 Example

In the case $U_1(c) = U_2(x) = \log c$, the functions of (7.10.9), (7.10.11) and (7.10.13) are given by

$$\mathcal{X}(y) = \frac{\alpha}{y} \, , \quad G(y) = -\alpha \cdot \log y + \delta \; ; \qquad 0 < y < \infty \qquad \text{and}$$
$$V(x) = \alpha \cdot \log(\frac{x}{\alpha}) + \delta \; ; \qquad 0 < x < \infty \, ,$$

where $\alpha \triangleq \alpha_1 + \alpha_2$, $\delta \triangleq \delta_1 + \delta_2$ in the notation of (7.8.19) and (7.9.16).

7.11 THE CASE OF CONSTANT COEFFICIENTS

The theory developed in the last three sections, culminating with Theorem 7.10.1, provides a very precise characterization of the value function for the optimization problem (7.10.2) (cf. expression (7.10.13)), as well as explicit formulas for the optimal processes of consumption rate $\hat{c}$ and wealth $\hat{X}$ (in (7.10.14), (7.10.15) respectively). But for the optimal portfolio process $\hat{\pi}$, the "martingale methodology" that we have employed so far is able to ascertain only its existence; there is no constructive algorithm, or a useful characterization, that could lead to its computation.

Our intent in the present section is to improve this situation; we shall impose Draconian assumptions on the model, which will enable us in particular to obtain the optimal $\hat{\pi}, \hat{c}$ in a very explicit *feedback form on the current level of wealth* (cf. (7.11.23), (7.11.24)).

Specifically, we shall assume throughout this section that

$$\beta(t) \equiv \beta, \quad r(t) \equiv r, \quad b(t) \equiv b, \quad \sigma(t) \equiv \sigma \ ; \qquad \forall \ t \in [0,T] \tag{7.11.1}$$

for given $\beta \in \mathcal{R}$, $r \in \mathcal{R}$, $b \in \mathcal{R}^d$, and σ a nonsingular, $(d \times d)$ matrix. This assumption will allow us to use "Markovian" methods, such as the Feynman-Kac representations of solutions to partial differential equations and the Hamilton-Jacobi-Bellman (HJB) equation of Dynamic Programming.

In order to make these methodologies available to us, we shall need a temporal as well as spatial parametrization; to wit, we write the analogues of the wealth equation (7.3.2) and of the value function (7.10.2) on the horizon $[t,T]$, for arbitrary $0 \le t \le T$, as

$$\begin{aligned} X(s) = x + \int_t^s (rX(u) - c(u))du + \int_t^s \pi^*(u)(b - r\underset{\sim}{1})du \\ + \int_t^s \pi^*(u)\sigma dW(u) \ ; \quad t \le s \le T \end{aligned} \tag{7.11.2}$$

and

$$V(t,x) = \sup_{(\pi,c) \ \in \ \mathcal{A}(t,x)} E\left[\int_t^T e^{-\beta s} U_1(c(s))ds + e^{-\beta T} U_2(X(T))\right] , \tag{7.11.3}$$

respectively. We shall also impose the purely technical assumptions

$$U_i(0) > -\infty \ , \quad \lim_{c \downarrow 0} \frac{(U_i'(c))^2}{U_i''(c))} \ \text{ exists}, \qquad \lim_{c \to \infty} \frac{(U_i'(c))^\alpha}{U_i''(c)} = 0 \ ; \quad i = 1,2 \tag{7.11.4}$$

for some $\alpha > 2$. They will permit the analysis to go through conveniently, and include as special cases the so-called HARA functions of the type $U(c) = (c+\eta)^\delta$; $0 < \delta <$

1, $\eta \geq 0$. However, they are far from being the weakest possible conditions under which the fundamental results will hold.

By analogy with our previous analysis and notation, let us introduce the vector $\theta = \sigma^{-1}(b - r\underset{\sim}{1}) \in \mathcal{R}^d$, the processes

$$Z_s^{(t)} \triangleq \exp\{-\theta^*(W_s - W_t) - \frac{1}{2}||\theta||^2(s-t)\}, \qquad \zeta_s^{(t)} \triangleq e^{(\beta - r)(s-t)} Z_s^{(t)} \qquad \text{and}$$

$$Y_s^{(t,y)} \triangleq y\zeta_s^{(t)} \; ; \quad t \leq s \leq T, \quad 0 < y < \infty \, ,$$

as well as the functions

$$G(t,y) \triangleq E[\int_t^T e^{-\beta(s-t)} U_1(I_1(Y_s^{(t,y)}))ds + e^{-\beta(T-t)} U_2(I_2(Y_T^{(t,y)}))] \tag{7.11.5}$$

$$\mathcal{X}(t,y) \triangleq E[\int_t^T e^{-\beta(s-t)} \zeta_s^{(t)} I_1(y\zeta_s^{(t)})ds + e^{-\beta(T-t)} \zeta_T^{(t)} I_2(y\zeta_T^{(t)})] \tag{7.11.6}$$

$$S(t,y) \triangleq y\mathcal{X}(t,y) = E[\int_t^T e^{-\beta(s-t)} Y_s^{(t,y)} I_1(Y_s^{(t,y)})ds + e^{-\beta(T-t)} Y_T^{(t,y)} I_2(Y_T^{(t,y)})] \tag{7.11.7}$$

for $(t,y) \in [0,T] \times (0,\infty)$. To avoid trivialities, we suppose $\theta \neq 0$; then for every $t \in [0,T)$ the function $\mathcal{X}(t,\cdot)$ is continuous and strictly decreasing on$(0,\infty)$, with $\mathcal{X}(t,\infty) = 0$. We denote its inverse by $\mathcal{Y}(t,\cdot)$, i.e.,

$$\mathcal{Y}(t, \mathcal{X}(t,y)) = y \; ; \qquad 0 \leq t < T, \;\; 0 \leq y \leq \infty$$

and by analogy with the characterization (7.10.13) of Theorem 7.10.1 we have

$$V(t,x) = e^{-\beta t} G(t, \mathcal{Y}(t,x)) \; ; \;\; (t,x) \in [0,T] \times (0,\infty) \, . \tag{7.11.8}$$

The point here is that we have reduced the study of the control problem (7.11.3) (or equivalently, of the nonlinear HJB equation (7.11.19) below, which is associated with it) to the study of the functions G, S of (7.11.5), (7.11.7); because, once these two are known, then $\mathcal{X}(t,y)$ is obtained straightaway as $y^{-1}S(t,y)$ and the value function V becomes available from (7.11.8). Now from the Feynman-Kac theorem, the functions G and S are characterized uniquely in terms of the Cauchy problems

$$(\frac{\partial}{\partial t} + L)G(t,y) + U_1(I_1(y)) = 0 \; ; \qquad (t,y) \in [0,T) \times (0,\infty) \tag{7.11.9}$$

$$G(T,y) = U_2(I_2(y)) \; ; \qquad y \in (0,\infty) \tag{7.11.10}$$

and

$$(\frac{\partial}{\partial t} + L)S(t,y) + yI_1(y) = 0 \; ; \qquad (t,y) \in [0,T) \times (0,\infty) \tag{7.11.11}$$

$$S(T,y) = yI_2(y) \; ; \qquad y \in (0,\infty) \; , \tag{7.11.12}$$

respectively, for the *linear* differential operator

$$L\varphi(t,y) \triangleq \frac{1}{2}||\theta||^2 y^2 \frac{\partial^2 \varphi(t,y)}{\partial y^2} + (\beta - r)y\frac{\partial \varphi(t,y)}{\partial y} - \beta\varphi(t,y) \; .$$

Indeed, using the conditions (11.4) it can be shown that G, S satisfy growth conditions of the type

$$\max_{0 \le t \le T} |u(t,y)| \le K(1 + y^\alpha + y^{-\alpha}) \; ; \qquad 0 < y < \infty \tag{7.11.13}$$

for some positive constants α, K, and that among such functions they are the unique solutions of their respective Cauchy problems. *We shall show how to compute these solutions in closed form* (Proposition 7.11.1).

To this end, let us recall Remark 7.6.7, and observe that the unique solution to the auxiliary Cauchy problem

$$(\frac{\partial}{\partial t} + L)v(t,y;\xi) = 0 \; ; \qquad (t,y) \in [0,T] \times (0,\infty)$$
$$v(T,y;\xi) = (\xi - y)^+ \; ; \qquad y \in (0,\infty)$$

is given by the Black & Scholes - type formula for a "put"option (the right to *sell* one share of the stock at the pre-assigned price $\xi > 0$:

$$v(t,y;\xi) = E[e^{-\beta(T-t)}(\xi - Y_T^{(t,y)})^+] \tag{7.11.14}$$

$$= \left\{ \begin{array}{lcl} \xi e^{-\beta(T-t)}\Phi(-\nu_-(T-t,y;\xi)) - ye^{-r(T-t)}\Phi(-\nu_+(T-t,y;\xi)) & ; & 0 \le t < T \\ (\xi - y)^+ & ; & t = T \end{array} \right\}$$

for every $(y,\xi) \in (0,\infty)^2$, where $\nu_\pm(t,y;\xi) \triangleq \frac{1}{\sqrt{2\gamma t}}[\log(\frac{y}{\xi}) + t(\beta - r \pm \gamma)]$ and $\gamma \triangleq ||\theta||^2/2$. Let us also introduce the functions

$$g(y) \triangleq \frac{U_1(I_1(y))}{\beta} - \frac{1}{\gamma(\lambda_+ - \lambda_-)}\left\{\frac{y^{1+\lambda_+}}{1+\lambda_+}J_+(y) - \frac{y^{1+\lambda_-}}{1+\lambda_-}\,J_-(y)\right\} \tag{7.11.15}$$

$$s(y) \triangleq \frac{yI_1(y)}{r} - \frac{1}{\gamma(\lambda_+ - \lambda_-)}\left\{\frac{y^{1+\lambda_+}}{\lambda_+} J_+(y) - \frac{y^{1+\lambda_-}}{\lambda_-} J_-(y)\right\}, \tag{7.11.16}$$

where $\lambda_+ > 0$ and $\lambda_- < 0$ are the roots of the quadratic equation

$$\gamma\lambda^2 - (r - \beta - \gamma)\lambda - r = 0$$

and

$$J_+(y) = \int_0^{I_1(y)} (U_1'(c))^{-\lambda_+} dc\ , \qquad J_-(y) = \int_1^{I_1(y)} (U_1'(c))^{-\lambda_-} dc\ .$$

It is easy to verify that g, s solve the ordinary differential equations

$$Lg(y) + U_1(I_1(y)) = 0\ , \qquad Ls(y) + yI_1(y) = 0\ ,$$

respectively.

One can now put the various results together, to arrive at the promised closed-form solutions.

7.11.1 Proposition

The functions G, S of (7.11.5), (7.11.7) have the stochastic representations

$$G(t,y) = g(y) + E[e^{-\beta(T-t)}\{U_2(I_2(Y_T^{(t,y)})) - g(Y_T^{(t,y)})\}]$$
$$S(t,y) = s(y) + E[e^{-\beta(T-t)}\{Y_T^{(t,y)} I_2(Y_T^{(t,y)})) - s(Y_T^{(t,y)})\}]\ ,$$

which lead to the closed-form expressions

$$G(t,y) = g(y) + (U_2(0) - \frac{U_1(0)}{\beta})e^{-\beta(T-t)} + \int_0^\infty (U_2(I_2(\xi)) - g(\xi))'' v(t,y;\xi)d\xi \tag{7.11.17}$$

$$S(t,y) = s(y) + \int_0^\infty (\xi I_2(\xi) - s(\xi))'' v(t,y;\xi)d\xi\ . \tag{7.11.18}$$

Moreover, the function $V : [0,T] \times [0,\infty) \to \mathcal{R}$ of (7.11.8) satisfies the *HJB Equation of Dynamic Programming*

$$\max_{\substack{c \geq 0 \\ \pi \in \mathcal{R}^d}} [\frac{1}{2}||\pi^*\sigma||^2 V_{xx}(t,x) + \{(rx - c) + \pi^*(b - r\underset{\sim}{1})\}V_x(t,x) + e^{-\beta t}U_1(c)]$$
$$= -V_t(t,x)\ ; \quad \text{in} \quad [0,T) \times (0,\infty) \tag{7.11.19}$$

and the terminal-boundary conditions

$$V(T,x) = e^{-\beta T}U_2(x)\ ; \quad 0 \leq x < \infty \tag{7.11.20}$$

$$V(t,0) = (U_2(0) - \frac{U_1(0)}{\beta})e^{-\beta T} + \frac{U_1(0)}{\beta}e^{-\beta t}\ ; \quad 0 \le t \le T\ , \tag{7.11.21}$$

respectively.

It is noteworthy that we have obtained a closed-form solution for the *nonlinear* HJB equation (7.11.19), by solving instead the two *linear* equations (7.11.9), (7.11.11) subject to the appropriate terminal and growth conditions, and then performing the composition (7.11.8).

The maximizations over $c \ge 0$, $\pi \in \mathcal{R}^d$ in (7.11.19) are achieved by

$$\hat{c} = I_1(\mathcal{Y}(t,x)), \qquad \hat{\pi} = -(\sigma\sigma^*)^{-1}(b - r\underset{\sim}{1})\frac{\mathcal{Y}(t,x)}{\mathcal{Y}_x(t,x)}\ .$$

This suggests that we should be able to justify similar *feedback form expressions* (on the current level of wealth) *for the optimal consumption and portfolio processes.* We choose to do this by studying directly the optimal wealth process.

7.11.2 Proposition

The optimal wealth process $X^{(t,x)}$ for the problem (7.11.3) is given by

$$X^{(t,x)}(s) = \mathcal{X}(s, \eta_s^{(t,x)})\ ; \quad t \le s \le T\ , \tag{7.11.22}$$

where $\eta_s^{(t,x)} \triangleq \mathcal{Y}(t,x)\zeta_s^{(t)} = Y_s^{(t,\mathcal{Y}(t,x))}$. In terms of it, the optimal pair $(\pi^{(t,x)},\ c^{(t,x)})$ is expressible as

$$c^{(t,x)}(s) = I_1(\mathcal{Y}(s, X^{(t,x)}(s))) \tag{7.11.23}$$

$$\pi^{(t,x)}(s) = -(\sigma\sigma^*)^{-1}(b - r\underset{\sim}{1})\ \frac{\mathcal{Y}(s, X^{(t,x)}(s))}{\mathcal{Y}_x(s, X^{(t,x)}(s))} \tag{7.11.24}$$

for $t \le s \le T$.

Proof: By analogy with (7.10.15), the optimal wealth process is given a.s. by

$$X^{(t,x)}(s) = E[\int_s^T e^{-r(\theta-s)} Z_\theta^{(s)} I_1(\eta_\theta^{(t,x)})d\theta + e^{-r(T-s)} Z_T^{(s)} I_2(\eta_T^{(t,x)})|\mathcal{F}_s]$$

$$= \frac{1}{Y_s^{(t,y)}}E[\int_s^T e^{-\beta(\theta-s)} Y_\theta^{(t,y)} I_1(Y_\theta^{(t,y)})d\theta + e^{-\beta(T-t)} Y_T^{(t,y)} I_2(Y_T^{(t,y)})|\mathcal{F}_s]$$

$$= \frac{S(s, Y_s^{(t,y)})}{Y_s^{(t,y)}}$$

with $y = \mathcal{Y}(t,x)$, for every $s \in [t,T]$; but this is (7.11.22). Now by analogy with (7.8.8), $\eta^{(t,x)}$ satisfies the equation

$$d\eta_s^{(t,x)} = \eta_s^{(t,x)}[(\beta - r)ds - \theta^* dW(s)] \ , \quad \eta_t^{(t,x)} = x \ . \tag{7.11.25}$$

On the other hand, by substituting $S(t,y) = y\mathcal{X}(t,y)$ into (7.11.11), one arrives at the linear parabolic equation

$$\mathcal{X}_t + \gamma y^2 \mathcal{X}_{yy} + (\beta - r + 2\gamma) y \mathcal{X}_y - r\mathcal{X} + I_1(y) = 0 \ ; \quad 0 \le t < T, \ 0 < y < \infty \tag{7.11.26}$$

for $\mathcal{X}(t,y)$. An application of Itô's rule to (7.11.22), in conjunction with (7.11.25) and (7.11.26), leads to

$$dX^{(t,x)}(s) = (rX^{(t,x)}(s) - c^{(t,x)}(s))ds + (\pi^{(t,x)}(s))^*[(b - r\underset{\sim}{1})ds + \sigma dW(s)]$$

in the notation of (7.11.23), (7.11.24). But this shows that $X^{(t,x)}$ is the wealth associated with the pair $(\pi^{(t,x)}, c^{(t,x)})$.

In the special case $U_1(c) = U_2(c) = c^\delta$ for some $0 < \delta < 1$, we have

$$G(t,y) = p(t)(\frac{y}{\delta})^{\frac{\delta}{\delta-1}}, \quad S(t,y) = \delta G(t,y), \quad \mathcal{X}(t,y) = p(t)(\frac{y}{\delta})^{\frac{1}{\delta-1}}$$

and

$$V(t,x) = e^{-\beta t}(p(t))^{1-\delta} x^\delta \ ,$$

as well as

$$c^{(t,x)}(s) = \frac{X^{(t,x)}(s)}{p(s)} \ , \quad \pi^{(t,x)}(s) = (\sigma\sigma^*)^{-1}(b - r\underset{\sim}{1})\frac{X^{(t,x)}(s)}{1-\delta} \ ; \quad t \le s \le T \ ,$$

where

$$p(t) = \begin{Bmatrix} \frac{1}{k}[1 - e^{-k(T-t)}] + e^{-k(T-t)} & ; \quad k \ne 0 \\ 1 + T - t & ; \quad k = 0 \end{Bmatrix} \ , \quad k = \frac{1}{1-\delta}(\beta - r\delta - \frac{\gamma\delta}{1-\delta}) \ .$$

7.12 THE PRICING OF AMERICAN OPTIONS

For the purposes of this section only, we shall need to modify slightly the model of section 7.3 for the small investor. First of all, we will have to deal with cumulative consumptions C_t up to time t, rather than with consumption rate processes.

7.12.1 Definition

A *consumption process* $C = \{C_t;\ 0 \le t \le T\}$ is continuous, increasing, adapted to $\{\mathcal{F}_t\}$, and satisfies $C_0 = 0,\ C_T < \infty$, almost surely.

Secondly, we have to allow the possibility for the stocks to pay *dividends* to the stockholders, at the rate $\mu_j(t);\ 0 \le t \le T$ for every dollar invested in the jth stock, $j = 1, \ldots, d$. These are nonnegative, bounded, measurable and $\{\mathcal{F}_t\}$-adapted processes, and we denote by $\mu(t) = (\mu_1(t), \ldots, \mu_d(t))^*$ the resulting vector process.

Then the wealth process X associated to a portfolio process π (Definition 7.3.1) and a consumption process C (Definition 7.12.1) satisfies the following analogue of equation (7.3.2):

$$\begin{aligned} dX(t) &= r(t)X(t)dt - dC_t + \pi^*(t)[b(t) + \mu(t) - r(t)\underset{\sim}{1}]dt + \pi^*(t)\sigma(t)dW(t) \\ &= r(t)X(t)dt - dC_t + \pi^*(t)\sigma(t)d\tilde{W}(t) \end{aligned} \tag{7.12.1}$$

in the notation of (7.4.2)-(7.4.4), with (7.4.1) replaced by

$$\theta(t) \triangleq (\sigma(t))^{-1}[b(t) + \mu(t) - r(t)\underset{\sim}{1}]. \tag{7.4.1}$$

The notion of admissibility for a pair (π, C) remains the same as in Definition 7.5.1, and (7.5.1) becomes:

$$\tilde{E}[X(\tau)e^{-\int_0^\tau r(s)ds} + \int_0^\tau e^{-\int_0^s r(u)du} dC_s] \le x\ ; \quad \forall\ \tau \in \mathcal{S}_{0,T} \tag{7.12.2}$$

for every $(\pi, C) \in \mathcal{A}(x)$.

After this setting of the stage, let us introduce the primary object of this section.

7.12.2 Definition

An *American Contingent Claim* (ACC) is a financial instrument consisting of

(i) an expiration date $T \in (0, \infty)$,

(ii) the selection of a stopping time $\tau \in \mathcal{S}_{0,T}$, and

(iii) a payoff $f(\tau)$ on exercise.

Here, $\{f(t);\ 0 \le t \le T\}$ is a continuous, nonnegative process, adapted to $\{\mathcal{F}_t\}$, which satisfies

$$E(\sup_{0 \le t \le T} f(t))^{\mu} < \infty , \qquad \text{for some} \qquad \mu > 1 . \tag{7.12.3}$$

For instance, if $f(t) = (P_1(t) - q)^+$, we have an *American option* on the 1st stock with exercise price $q \ge 0$, exercisable at any stopping time τ on $[0,T]$. We restrict attention to stopping times, in order to exclude clairvoyance and/or insider trading!

As in section 7.6, we are interested in the following *pricing problem* for the ACC: What is a fair price to pay at $t = 0$ for this instrument ? How much is it worth at any later time $t \in (0,T]$?

Let us suppose for a moment that the selection of $\tau \in \mathcal{S}_{0,T}$ ((ii) in Definition 7.12.2)) has been made; then we have, from Remark 7.6.8, the expressions

$$X^{(\tau)}(t) = \tilde{E}[f(\tau)e^{-\int_t^\tau r(s)ds}|\mathcal{F}_t] , \qquad v^{(\tau)} = X^{(\tau)}(0) = \tilde{E}[f(\tau)e^{-\int_0^\tau r(s)ds}]$$

for the value of the claim and for its fair price at $t = 0$. It is conceivable then that, in order to find the corresponding quantities for the ACC, one would merely have to maximize over stopping times. In particular, if we should expect the fair price at $t = 0$ to be given by

$$\sup_{\tau \in \mathcal{S}_{0,T}} \tilde{E}[f(\tau)e^{-\int_0^\tau r(s)ds}],$$

and the value of the ACC at any time $t \in [0,T]$ by

$$\text{ess sup}_{\tau \in \mathcal{S}_{t,T}} \tilde{E}[f(\tau)e^{-\int_t^\tau r(s)ds}|\mathcal{F}_t] , \quad \text{a.s.}$$

The question is whether this process is the wealth corresponding to an admissible portfolio/consumption process pair, that somehow again *duplicates* the payoff from the contingent claim and does so with minimal initial endowment.

7.12.3 Definition

Fix $x > 0$; a pair $(\pi, X) \in \mathcal{A}(x)$ is called a *hedging strategy* against the ACC with initial wealth x, if the corresponding wealth process X satisfies

(i) $X(t) \ge f(t)$; $\quad \forall \ 0 \le t \le T$

(ii) $X(T) = f(T)$,

almost surely. We denote by $\hat{\mathcal{H}}(x)$ the collection of all such pairs.

7.12.4 Definition

The number

$$\hat{v} \triangleq \inf\{x > 0 \; ; \; \exists \, (\pi, C) \in \hat{\mathcal{H}}(x)\} \tag{7.12.4}$$

is called the *fair price* for the ACC of Definition 7.12.2.

For every $(\pi, C) \in \hat{\mathcal{H}}(x)$, we have from (7.12.2):

$$\tilde{E}[f(\tau)e^{-\int_0^\tau r(s)ds}] \leq x \, , \quad \forall \; \tau \in \mathcal{S}_{0,T}.$$

Therefore, with

$$u(t) \triangleq \sup_{\tau \in \mathcal{S}_{t,T}} \tilde{E}Q(\tau) \, , \quad Q(t) = f(t)e^{-\int_0^t r(s)ds} \, , \quad 0 \leq t \leq T \, , \tag{7.12.5}$$

we obtain $u(0) \leq x$, whence

$$u(0) \leq \hat{v} \, . \tag{7.12.6}$$

We shall show that equality actually holds in (7.12.6).

7.12.5 Theorem

The fair price $\hat{v}$ of Definition 7.12.4 is given by

$$\hat{v} = u(0) = \sup_{\tau \in \mathcal{S}_{0,T}} \tilde{E}[f(\tau)e^{-\int_0^\tau r(s)ds}] \, . \tag{7.12.7}$$

There exists a pair $(\hat{\pi}, \hat{C}) \in \hat{\mathcal{H}}(u(0))$, such that the corresponding wealth process $\hat{X}$ is given as

$$\hat{X}(t) = \text{ess sup}_{\tau \in \mathcal{S}_{t,T}} \tilde{E}[f(\tau)e^{-\int_t^\tau r(s)ds} | \mathcal{F}_t] \, , \quad \text{a.s.} \tag{7.12.8}$$

for every $t \in [0, T]$, and

$$\int_0^T 1_{\{X(t) > f(t)\}} \, d\hat{C}_t = 0 \, , \quad \text{a.s.} \tag{7.12.9}$$

holds.

In view of (7.12.6), only the second claim needs to be discussed. In order to do this, we have to recall some basic facts from the *theory of optimal stopping* for a continuous process such as Q (cf. Fakeev (1970), Bismut & Skalli (1977), El Karoui (1981)). We know from these sources that there exists a nonnegative, RCLL (**R**ight-**C**ontinuous with **L**eft-hand **L**imits) $\tilde{P}$-supermartingale $\{Y(t), \mathcal{F}_t;\ 0 \le t \le T\}$, such that the function $u(\cdot)$ of (7.12.5) is given as

$$u(t) = \tilde{E}Y(t)\ ; \qquad 0 \le t \le T\ , \tag{7.12.10}$$

and

$$Y(t) = \text{ess sup}_{\tau \epsilon \mathcal{S}_{t,T}}\ \tilde{E}[Q(\tau)|\mathcal{F}_t]\ , \qquad \text{a.s.} \tag{7.12.11}$$

holds for every $t \epsilon [0, T]$. Y is the *Snell envelope* of Q, i.e., the smallest RCLL supermartingale that dominates Q, and provides the optimal stopping time τ_t for the problem of (7.12.5): $u(t) = \tilde{E}Q(\tau_t)$, in the form

$$\tau_t \triangleq \inf\{s \epsilon [t, T]\ ;\ \ Y(s) = Q(s)\}\ . \tag{7.12.12}$$

Using (7.12.3) and the Doob and Jensen inequalities, it can be shown that Y *is of the class* $D[0, T]$ *under* $\tilde{P}$, i.e., that

$$\{Y(\tau)\}_{\tau \epsilon \mathcal{S}_{0,T}} \ \text{ is a } \ \tilde{P} - \text{uniformly integrable family.} \tag{7.12.13}$$

Bismut & Skalli (1977) also show that Y *is regular*:

$$\left\{ \begin{array}{c} \text{for every monotone sequence } \{\sigma_n\}_{n=1}^{\infty} \subseteq \mathcal{S}_{0,T} \text{ with} \\ \lim_{n\to\infty} \sigma_n = \sigma \ \epsilon\ \mathcal{S}_{0,T}\ , \quad \text{we have}: \ \lim_{n\to\infty} \tilde{E}Y(\sigma_n) = \tilde{E}Y(\sigma) \end{array} \right. \tag{7.12.14}$$

Proof of Theorem 7.12.5

From (7.12.13), (7.12.14) we conclude that Y admits the Doob-Meyer decomposition (e.g. Karatzas & Shreve (1987), Section 7.1.4):

$$Y(t) = u(0) + M_t - A_t\ ; \qquad 0 \le t \le T\ ,$$

where $\{M_t, \mathcal{F}_t\}$ is a $\tilde{P}$-martingale and A is a *continuous*, nondecreasing process, with $M_0 = A_0 = 0$. As in the proof of Theorem 7.5.3, we have the representation

$$M_t = \int_0^t e^{-\int_0^s r(u)du}\ \hat{\pi}^*(s)\sigma(s)d\tilde{W}(s)\ ; \qquad 0 \le t \le T$$

of the martingale M as a stochastic integral with respect to $\tilde{W}$, for a suitable portfolio process $\hat{\pi}$. Now define

$$\hat{X}(t) \triangleq Y(t)e^{\int_0^t r(s)ds} ; \qquad 0 \le t \le T , \tag{7.12.15}$$

and apply Itô's rule to obtain

$$d\hat{X}(t) = r(t)\hat{X}(t)dt - d\hat{C}_t + \hat{\pi}^*(t)\sigma(t)d\tilde{W}(t)$$

for the choice

$$\hat{C}_t \triangleq \int_0^t e^{-\int_0^s r(u)du} \, dA_s . \tag{7.12.16}$$

In other words, $\hat{X}$ is the wealth process corresponding to the portfolio/consumption process pair $(\hat{\pi}, \hat{C})$, which is easily seen to belong to $\hat{\mathcal{H}}(u(0))$. The representation (7.12.8) follows from (7.12.11), and (7.12.9) from

$$\int_0^T 1_{\{Y(t)>Q(t)\}} \, dA_t = 0 , \qquad \text{a.s.}$$

(cf. Bismut & Skalli (1977), El Karoui (1981)).

The stopping time τ_t of (7.12.12) can be written equivalently as

$$\tau_t = \inf\{s \in [t,T] ; \quad X(s) = f(s)\} , \quad \text{a.s.} ; \tag{7.12.17}$$

obviously, τ_o *provides the optimal exercise time for the ACC.* The random variable $\hat{X}(t)$ of (7.12.8) gives the value of the ACC at time t.

7.12.6 Remark

Suppose that the process Q of (7.12.5) is a submartingale under $\tilde{P}$; then $u(t) = \tilde{E}Q(T)$, $\tau_t = T$ for every $0 \le t \le T$, and *the ACC should not be exercised before the expiration date* (i.e., is equivalent to a ECC).

For instance, in the setting of Example 7.6.6, suppose that the function $\varphi : \mathcal{R}_+^d \to [0,\infty)$ is of class C^2 and satisfies

$$\frac{1}{2}\sum_{j=1}^d \sum_{k=1}^d a_{jk}p_j p_k \frac{\partial^2 \varphi(p)}{\partial p_j \partial p_k} + \sum_{j=1}^d (r - \mu_j)p_j \frac{\partial \varphi(p)}{\partial p_j} \ge r\varphi(p)$$

as well as a polynomial growth condition. Then $Q(t) = e^{-rt}\varphi(P(t))$ is a $\tilde{P}$-submartingale, and the value process for the ACC with $f(t) = \varphi(P(t))$ is given by (7.6.12), with the understanding that r has to be replaced by $r - \mu_j$ in the expressions (7.6.7) - (7.6.9).

As another example, take the American option with $d = 1$, $f(t) = (P_1(t) - q)^+$, $q > 0$ written on a stock which pays no dividends: $\mu_1(t) \equiv 0$, $r(t) \geq 0$. Then

$$Q(t) = (P_1(t)e^{-\int_0^t r(s)ds} - qe^{-\int_0^t r(s)ds})^+$$

is a $\tilde{P}$-submartingale, and we recover a result of Merton (1973): *an American option with positive exercise price, written on a stock that pays no dividends, should not be exercised before the expiration date.*

7.12.7 The infinite horizon case

In the setting of Example 7.6.6 and with $\mu_i(t) \equiv \mu$, $f(t) = \varphi(P(t))$, the value process $\hat{X}$ of (7.12.8) for $T = \infty$ is given, formally at least, as

$$\hat{X}(t) = v(P(t)) \; ; \qquad 0 \leq t < \infty \, , \tag{7.12.18}$$

where $v : \mathcal{R}_+^d \to [0, \infty)$ is the least r-excessive majorant of the function φ (Fakeev (1971)).

More specifically, if $d = 1$ and $\varphi(p) = (p-1)^+$, the function v of (7.12.18) was computed by Mc Kean (1965) as

$$v(p) = \left\{ \begin{matrix} (\kappa - 1)(\frac{p}{\kappa})^\gamma & ; & 0 < p < \kappa \\ p - 1 & ; & \kappa \leq p < \infty \end{matrix} \right\}$$

with $\gamma = \frac{1}{\sigma^2}\left(\sqrt{\delta^2 + 2r\sigma^2} - \delta\right)$, $\alpha = r - \mu > 0$, $\delta = \alpha - \frac{\sigma^2}{2}$, $\kappa = \frac{\gamma}{\gamma - 1} > 1$, and the optimal exercise time of (7.12.17) is given by

$$\tau_t = \inf\{s \geq t \; ; \quad P_1(s) \geq \kappa\} \, .$$

The finite-horizon version of this problem is studied in Van Moerbeke (1976); one faces then a genuine free-boundary problem, for a moving boundary $\kappa(t)$; $0 \leq t \leq T$ rather than a point κ as above, which Van Moerbeke studies by reducing it to a problem of the Stefan type.

7.13 AN EQUILIBRIUM MODEL

Let us consider in this final section an *economy*, which consists of

(a) the same financial market as in section 7.2,

(b) a single "commodity" which is traded at the *spot price* $\psi = \{\psi(t); 0 \le t \le T\}$, and

(c) a finite number n of agents (small investors). Each one of these receives a *commodity endowment* at the rate $e_i = \{e_i(t);\ 0 \le t \le T\}$, invests in the financial market, and consumes the commodity at a rate that maximizes his expected discounted utility from consumption (as in section 7.8).

The *equilibrium problem* for such an economy, is to determine a spot price ψ, so that the markets clear when each agent behaves optimally and the commodity is traded at the price ψ. We shall show that the methodology of section 7.8 is ideally suited to handle this question.

We formulate now the problem a bit more carefully. The primitives in this economy are the financial market of section 7.2 and the commodity endowment processes $e_1, \dots, e_n$; these are assumed to be nonnegative, measurable and adapted to the filtration $\{\mathcal{F}_t\}$.

7.13.1 Definition

A process $\psi = \{\psi(t);\ 0 \le t \le T\}$ with values in $(0, \infty)$ is an *admissible spot price for the endowments* $e_1, \dots, e_n$, if it is measurable, adapted to $\{\mathcal{F}_t\}$, and satisfies

$$x_i \triangleq \tilde{E} \int_0^T e^{-\int_0^t r(s)ds}\, \psi(t) e_i(t) dt < \infty\ ; \quad \forall\ i = 1, \dots, n\ . \tag{7.13.1}$$

Given such a spot price process ψ, the ith agent has at his disposal the choice of a *portfolio* process $\pi_i(t) = (\pi_{i1}(t), \dots, \pi_{id}(t))^*$ and of a *consumption rate* process $c_i(t);\ 0 \le t \le T$. These are as in Definitions 7.3.1 and 7.3.2, except now (7.3.1) is replaced by

$$\int_0^T \psi(t) c_i(t) dt < \infty\ , \qquad \text{a.s.} \tag{7.3.1$'$}$$

For every such pair (π_i, c_i), the corresponding *wealth process* X_i satisfies, by analogy with (7.3.2), the equation

$$dX_i(t) = r(t)X_i(t)dt + \psi(t)[e_i(t) - c_i(t)]dt + \pi_i^*(t)[b(t) - r(t)\underset{\sim}{1}]dt + \pi_i^*(t)\sigma(t)dW(t) .$$

In terms of the $\tilde{P}$-Brownian motion $\tilde{W}$ of (7.4.4), the solution is given by

$$X_i(t)e^{-\int_0^t r(s)ds} = \int_0^t e^{-\int_0^s r(u)du} \psi(s)[e_i(s) - c_i(s)]ds + \int_0^t e^{-\int_0^s r(u)du} \pi_i^*(s)\sigma(s)d\tilde{W}(s) . \tag{7.13.2}$$

We shall need to replace the notion of admissibility for portfolio/consumption rate process pairs (Definition 7.5.1) by one with a less stringent nonnegativity requirement.

7.13.2 Definition

Given an admissible spot price process ψ, a portfolio/consumption process pair (π_i, c_i) is *feasible for the ith agent*, if the corresponding wealth process X_i of (7.13.2)

(a) is of class $D[0,T]$ under $\tilde{P}$, and

(b) satisfies $X_i(T) \geq 0$, a.s.

The optimization problem faced by the ith agent, is now to maximize the expected discounted utility from consumption

$$E\int_0^T e^{-\int_0^t \beta(s)ds} U_i(c_i(t))dt \tag{7.13.3}$$

over all feasible pairs (π_i, c_i), for a given utility function as in section 7.7. We shall let $(\hat{\pi}_i, \hat{c}_i)$ denote an optimal pair for this problem, and $\hat{X}_i$ the associated wealth process.

We are now in a position to define the notion of equilibrium for the economy.

7.13.3 Definition

An admissible spot price process ψ *supports an equilibrium*, if we have

(a) clearing of the spot markets, i.e.,

$$\sum_{i=1}^n \hat{c}_i(t) = \sum_{i=1}^n e_i(t) ; \qquad \forall \ 0 \leq t \leq T , \text{ and} \tag{7.13.4}$$

(b) clearing of the financial markets, i.e.,

$$\sum_{i=1}^{n} \hat{\pi}_{ij}(t) = 0 \; ; \qquad \forall \;\; 0 \le t \le T, \quad \forall \; j = 1, \ldots, d \tag{7.13.5}$$

$$\sum_{i=1}^{n} \hat{X}_i(t) = 0 \; ; \qquad \forall \;\; 0 \le t \le T \tag{7.13.6}$$

almost surely.

For a given ψ, let us try to solve the ith optimization problem, in the manner of section 7.8. First, let us state the analogue of (7.5.4) and Theorem 7.5.3.

7.13.4 Proposition

Let ψ be an admissible spot price process, and (π_i, c_i) be a feasible pair for the ith agent. Then in the notation of (7.13.1), (7.8.6) we have

$$\tilde{E} \int_0^T e^{-\int_0^t r(s)ds} \psi(t)c_i(t)dt = E \int_0^T e^{-\int_0^t \beta(s)ds} \zeta_t \psi(t)c_i(t)dt \le x_i \; . \tag{7.13.7}$$

Conversely, suppose c_i is a consumption rate process satisfying (7.13.7); then there exists a portfolio process π_i, such that the pair (π_i, c_i) is feasible for agent i.

Proof: Introduce the stopping times

$$\tau_k \triangleq \inf\{0 \le t \le T \; ; \; \int_0^t ||\pi_i^*(s)\sigma(s)||^2 ds \ge k\} \wedge T \; ; \qquad k = 1, 2, \ldots \; .$$

From (7.13.2) we obtain

$$\begin{aligned} \tilde{E} X_i(\tau_k) e^{-\int_0^{\tau_k} r(s)ds} &+ \tilde{E} \int_0^{\tau_k} e^{-\int_0^t r(s)ds} \psi(s)c_i(s)ds \\ &= \tilde{E} \int_0^{\tau_k} e^{-\int_0^t r(s)ds} \psi(s)e_i(s)ds \le x_i; \end{aligned}$$

now let $k \to \infty$, and use the monotone convergence theorem and the conditions of Definition 7.13.2, to conclude that (7.13.7) follows.

For the converse, one proceeds by analogy with the proof of Theorem 7.5.3 : consider the random variable

$$D_i \triangleq \int_0^T e^{-\int_0^s r(u)du} \psi(s)[e_i(s) - c_i(s)]ds \; , \tag{7.13.8}$$

the $\tilde{P}$-martingale

$$\begin{aligned} M_i(t) &\triangleq \tilde{E}D_i - \tilde{E}(D_i|\mathcal{F}_t) \\ &= \int_0^t e^{-\int_0^s r(u)du}\, \pi_i^*(s)\sigma(s)d\tilde{W}(s)\ ; \quad 0 \le t \le T \end{aligned} \tag{7.13.9}$$

(from the martingale representation theorem, for a suitable partfolio process π_i), and define X_i via

$$X_i(t)e^{-\int_0^t r(s)ds} \triangleq \int_0^t e^{-\int_0^s r(u)du}\, \psi(s)[e_i(s) - c_i(s)]ds + M_i(t)\ ; \quad 0 \le t \le T\ . \tag{7.13.10}$$

It is easily checked that $X_i \epsilon D[0,T]$, and

$$X_i(T) = \tilde{E}D_i \cdot e^{-\int_0^T r(s)ds} \ge 0\ ;$$

hence, X_i is the wealth process associated with the feasible pair (π_i, c_i).

It develops that the ith agent's optimization problem amounts to maximizing the expression (7.13.3), subject to the constraint (7.13.7). As in Theorem 7.8.1, the *optimal consumption rate process* is given by

$$\hat{c}_i(t) = I_i(y_i\psi(t)\zeta_t) \tag{7.13.11}$$

where the constant $y_i > 0$ is determined by the requirement that (7.13.7) hold as an identity; i.e.,

$$\begin{aligned} &E\int_0^T e^{-\int_0^t \beta(s)ds}\psi(t)\zeta_t . I_i(y_i\psi(t)\zeta_t)dt \\ &\qquad E\int_0^T e^{-\int_0^t \beta(s)ds}\, \psi(t)\zeta_t . e_i(t)dt\ ; \quad i = 1, \ldots, n\ . \end{aligned} \tag{7.13.12}$$

For this choice, the corresponding quantities $\hat{D}_i, \hat{M}_i, \hat{\pi}_i$ and $\hat{X}_i$ in (7.13.8) - (7.13.10) satisfy $\tilde{E}(\hat{D}_i) = 0$, $\hat{X}_i(T) = 0$.

Finally, with $\hat{e}(t) \triangleq \Sigma_{i=1}^n e_i(t)$ and taking (7.13.11) into account, the condition (7.13.4) amounts to

$$\sum_{i=1}^n I_i(y_i\psi(t)\zeta_t) = \hat{e}(t)\ ; \quad 0 \le t \le T \tag{7.13.13}$$

almost surely.

7.13.5 Proposition

Suppose there exist a vector $(y_1, \ldots, y_n) \in \mathcal{R}_+^n$ and an admissible spot price process ψ, such that (7.13.12), (7.13.13) are satisfied. Then ψ supports an equilibrium.

Proof: From (7.13.8), (7.13.13) we have $\Sigma_{i=1}^n \hat{D}_i = 0$ a.s., whence

$$\sum_{i=1}^n \hat{M}_i(t) = -\tilde{E}(\sum_{i=1}^n \hat{D}_i | \mathcal{F}_t) = 0, \qquad \sum_{i=1}^n \hat{X}_i(t) = 0\ ; \quad 0 \le t \le T$$

almost surely. Thus (7.13.4), (7.13.6) are verified, and it is not hard to see that the portfolios $\hat{\pi}_i\ ;\ 1 \le i \le n$ can also be chosen to satisfy (7.13.5).

Questions of existence and uniqueness for the system of (7.13.12), (7.13.13) will be addressed elsewhere. We shall content ourselves here with discussing two particular cases.

7.13.6 Example

Suppose $U_i(c) = \log c\ ;\ 1 \le i \le n$. Then

$$y_i = \left(\frac{E \int_0^T e^{-\int_0^t \beta(s)ds} \frac{e_i(t)}{\hat{e}(t)} dt}{E \int_0^T e^{-\int_0^t \beta(s)ds}\, dt} \right)^{-1} \ ; \quad 1 \le i \le n$$

provides the unique solution subject to $\sum_{i=1}^n y_i^{-1} = 1$; the corresponding spot price process is given by

$$\psi(t) = \frac{1}{\zeta_t . \hat{e}(t)} \, .$$

7.13.7 Example

If $U_i(c) = c^\delta\ ;\ 1 \le i \le n$ with $0 \le \delta < 1$,

$$y_i = \left(\frac{E \int_0^T e^{-\int_0^t \beta(s)ds} \frac{e_i(t)}{(\hat{e}(t))^{1-\delta}} dt}{E \int_0^T e^{-\int_0^t \beta(s)ds} (\hat{e}(t))^\delta dt} \right)^{\delta - 1} \ ; \quad 1 \le i \le n$$

gives the unique solution subject to $\Sigma_{i=1}^n y_i^{1/(\delta-1)} = 1$; the spot price process corresponding to this vector is

$$\psi(t) = \frac{1}{\zeta_t (\hat{e}(t))^{1-\delta}} \, .$$

7.14 NOTES

Sections 7.2-4 The idea of introducing a probability measure, under which the discounted stock prices of (7.4.7) are martingales, is due to Harrison and Kreps (1979) and Harrison and Pliska (1981, 1983). The model that we have adopted, with the particular completeness condition (7.2.3), is due to Bensoussan (1984).

Section 7.5 The material here comes from Karatzas and Shreve (1987), section 7.5.8 and Karatzas, Lehoczky and Shreve (1987). The terminology "attainable levels of terminal wealth" is due to Pliska (1986), who has a result similar to Theorem 7.5.4(ii).

Section 7.6 We follow Karatzas & Shreve (1987). Example 7.6.6 is adapted from Harrison & Pliska (1981). For comprehensive accounts of option pricing, see Samuelson (1973), Merton (1973) and Smith (1976). For a model of option pricing, in which the borrowing rate is higher than the interest rate of the bond, see Barron & Jensen (1987).

Sections 7.8-7.11 come from Karatzas, Lehoczky and Shreve (1987), which should be consulted for the details which are only sketched in these notes. The model with constant coefficients and utility from consumption, was introduced by Samuelson (1969) and Merton (1969, 1971) for utility functions of the HARA class, and was studied extensively by Karatzas, Lehoczky, Sethi and Shreve (1986) for general utility functions, using the HJB equation of dynamic programming and allowing for general patterns of behaviour upon bankruptcy.

Related results have been obtained independently by Cox and Huang (1986, 1987).

Recent work by Davis and Norman (1987) treats a model with constant coefficients, one stock and utility $U(c) = \sqrt{c}$ from consumption, but with *costs of transaction* between the two assets.

Section 7.12 is adapted from Karatzas (1987); see also Bensoussan (1984), for a different approach to the stopping problem.

Section 7.13: is taken from Lehoczky and Shreve (1987). See also Duffie (1986), Duffie and Huang (1985) for related equilibrium models, as well as Cox, Ingersoll and Ross (1985). In joint work with Lehoczky and Shreve, currently in progress, we extend the results of this section in the context of a model with endogenous determination of asset prices; we also discuss questions of existence and uniqueness for the corresponding equations (7.13.12), (7.13.13).

7.15 ACKNOWLEDGEMENTS

I would like to thank Prof. John Baras, for extending to me the invitation to deliver these lectures and for the hospitality and support of the Systems Research Center at the University of Maryland. In these seminars, I had the good fortune to find a very attentive and stimulating participant in the person of Prof. L.C. Evans.

I owe a lot to my audiences at the Université Pierre et Marie Curie (Paris VI) and at Columbia University, where I also lectured extensively on these matters. I am particularly indebted to Profs. Aline Bonami and Dominique Lépingle, for studying this material in their seminar at the Université d' Orléans and for sending me their notes, which in turn influenced the present exposition.

A great deal of this paper reports on joint work with Profs. John Lehoczky and Steven Shreve, carried out during the course of the last two years. I wish to extend my appreciation to John and Steve for our collaboration, and hope that they will be pleased with this attempt at reviewing our work.

REFERENCES

[1] BARRON, E. N. and JENSEN, R. (1987). *A stochastic control approach to the pricing of options.* Preprint.

[2] BENSOUSSAN, A. (1984). *On the theory of option pricing.* Acta Applicandae Mathematicae, 2, 139-158.

[3] BISMUT, J. M. and SKALLI, B. (1977). *Temps d'arrêt optimal, théorie générale de processus et processus de Markov.* Z. Wahrsch. verw. Geb. 39, 301-313.

[4] BLACK, F. and SCHOLES, M. (1973). *The pricing of options and corporate liabilities.* J. Political Economy, 81, 637-659.

[5] COX, J. C., INGERSOLL, J. and ROSS, S. (1985). *An intertemporal general equilibrium model of asset prices.* Econometrica 53, 363-384.

[6] COX, J. C. and HUANG, C. (1986). *A variational problem arising in financial economics with an application to a portfolio turnpike theorem.* Preprint.

[7] COX, J. C. and HUANG, C. (1987). *Optimal consumption and investment policies when asset prices follow a diffusion process.* Preprint.

[8] DAVIS, M. H. A. and NORMAN, A. R. (1987). *Portfolio selection with transaction costs.* Preliminary draft, March 1987.

[9] DUFFIE, D. (1986). *Stochastic equilibria: existence, spanning number, and the "no expected financial gain from trade"hypothesis.*

[10] DUFFIE, D. and HUANG, C. (1985). *Implementing Arrow-Debreu equilibria by continuous trading of few long-lived securities.* Econometrica 53, 1337-1356.

[11] EL KAROUI, N. (1981). *Les aspects probabilistes du contrôle stochastique.* Lecture Notes in Mathematics, 876, 73-238. Springer Verlag, Berlin.

[12] FAKEEV, A. G. (1970). *Optimal stopping rules for processes with continuous parameter.* Theor. Probab. Appl. 15, 324-331.

[13] FAKEEV, A. G. (1971). *Optimal stopping of a Markov process.* Theory Probab. Appl. 16, 694-696.

[14] GIRSANOV, I. V. (1960). *On transforming a certain class of stochastic processes by absolutely continuous substitution of measures.* Theory Probab. Appl. 5, 285-301.

[15] HARRISON, J. M. and KREPS, D. M. (1979). *Martingales and arbitrage in multiperiod security markets.* J. Economic Theory 20, 381-408.

[16] HARRISON, J. M. and PLISKA, S. R. (1981). *Martingales and stochastic integrals in the theory of continuous trading.* Stochastic Processes Appl. 11, 215-260.

[17] HARRISON, J. M. and PLISKA, S. R. (1983). *A stochastic calculus model of continuous trading: complete markets.* Stochastic Processes Appl. 15, 313-316.

[18] IKEDA, N. and WATANABE, S. (1981). *Stochastic Differential Equations and Diffusion Processes.* North-Holland, Amsterdam and Kodansha Ltd., Tokyo.

[19] KARATZAS, I. (1987). *On the pricing of American options.* Appl. Math. Optimization, to appear.

[20] KARATZAS, I., LEHOCZKY, J. P., SETHI, S. P. and SHREVE, S. E. (1986) *Explicit solution of a general consumption/investment problem.* Mathem. Operations Research 11, 261-294.

[21] KARATZAS, I., LEHOCZKY, J. P., SHREVE, S. E. (1987). *Optimal portfolio and consumption decisions for a "small investor"on a finite horizon.* SIAM J. Control & Optimization, to appear.

[22] KARATZAS, I. and SHREVE, S. E. (1987) *Brownian Motion and Stochastic Calculus.* Springer-Verlag, New York (in press).

[23] LEHOCZKY, J. P., and SHREVE, S. E. (1987). *Explicit equilibrium solutions for a multi-agent consumption/investment problem.* Technical Report 384, Department of Statistics, Carnegie Mellon University.

[24] McKEAN, H. P., Jr. (1965). Appendix to [29]: *A free boundary problem for the heat equation arising from a problem in mathematical economics.* Industr. Manag. Review 6, 32-39.

[25] MERTON, R. C. (1969). *Lifetime portfolio selection under uncertainty: continuous-time case*. Rev. Econom. Statist. 51, 247-257.

[26] MERTON, R. C. (1971). *Optimum consumption and portfolio rules in a continuous-time model.* J. Economic Theory 3, 373-413. Erratum: ibid. 6 (1973), 213-214.

[27] MERTON, R. C. (1973). *Theory of rational option pricing.* Bell J. Econom. Manag. Sci. 4, 141-183.

[28] PLISKA, S. R. (1986). *A stochastic calculus model of continuous trading: optimal portfolio.* Mathem. Operations Research 11, 371-382.

[29] SAMUELSON, P. A. (1965). *Rational theory of warrant pricing.* Industr. Manag. Review 6, 13-31.

[30] SAMUELSON, P. A. (1969). *Lifetime portfolio selection by dynamic stochastic programming.* Rev. Econom. Statist. 51, 239-246.

[31] SAMUELSON, P. A. (1973). *Mathematics of speculative prices.* SIAM Review 15, 1-42.

[32] SMITH, C. W., Jr. (1976). *Option pricing: a review.* J. Financial Economics 3, 3-51.

[33] VAN MOERBEKE, P. (1976). *On optimal stopping and free boundary problems.* Arch. Rational Mech. and Analysis 60, 101-148.

MIX
Papier aus verantwortungsvollen Quellen
Paper from responsible sources
FSC® C105338

If you have any concerns about our products,
you can contact us on
ProductSafety@springernature.com

In case Publisher is established outside the EU,
the EU authorized representative is:
Springer Nature Customer Service Center GmbH
Europaplatz 3, 69115 Heidelberg, Germany

Printed by Libri Plureos GmbH
in Hamburg, Germany